Sinusoidal Vibration

Mechanical Vibration & Shock

Sinusoidal Vibration

Volume I

Christian Lalanne

First published in 1999 by Hermes Science Publications, Paris
First published in English in 2002 by Hermes Penton Ltd

Taylor & Francis Books, Inc
29 West 35th Street
New York, NY 10001

© Hermes Science Publications, 1999
© English language edition Hermes Penton Ltd, 2002

ISBN 1-56032-985-8

Printed and bound in Great Britain by Biddles Ltd, Guildford and King's Lynn

Contents

Foreword to series

In the course of their lifetime simple items in everyday use such as mobile telephones, wristwatches, electronic components in cars or more specific items such as satellite equipment or flight systems in aircraft, can be subjected to various conditions of temperature and humidity and more especially to mechanical shock and vibrations, which form the subject of this work. They must therefore be designed in such a way that they can withstand the effect of the environment conditions they are exposed to without being damaged. Their design must be verified using a prototype or by calculations and/or significant laboratory testing.

Currently, sizing and testing are performed on the basis of specifications taken from national or international standards. The initial standards, drawn up in the 1940s, were blanket specifications, often extremely stringent, consisting of a sinusoidal vibration the frequency of which was set to the resonance of the equipment. They were essentially designed to demonstrate a certain standard resistance of the equipment, with the implicit hypothesis that if the equipment survived the particular environment it would withstand undamaged the vibrations to which it would be subjected in service. The evolution of those standards, occasionally with some delay related to a certain conservatism, followed that of the testing facilities: the possibility of producing swept sine tests, the production of narrow-band random vibrations swept over a wide range, and finally the generation of wide-band random vibrations. At the end of the 1970s, it was felt that there was a basic need to reduce the weight and cost of on-board equipment, and to produce specifications closer to the real conditions of use. This evolution was taken into account between 1980 and 1985 concerning American standards (MIL-STD 810), French standards (GAM EG 13) or international standards (NATO) which all recommended the *tailoring of tests*. Current preference is to talk of the *tailoring of the product to its environment* in order to assert more clearly that the environment must be taken into account as from the very start of the project, rather than to check the behaviour of the material *a posteriori*. These concepts, originating with the military, are currently being increasingly echoed in the civil field.

Tailoring is based on an analysis of the life profile of the equipment, on the measurement of the environmental conditions associated with each condition of use and on the synthesis of all the data into a simple specification, which should be of the same severity as the actual environment.

This approach presupposes a proper understanding of the mechanical systems subjected to dynamic loads, and knowledge of the most frequent failure modes.

Generally speaking, a good assessment of the stresses in a system subjected to vibration is possible only on the basis of a finite elements model and relatively complex calculations. Such calculations can only be undertaken at a relatively advanced stage of the project once the structure has been sufficiently defined for such a model to be established.

Considerable work on the environment must be performed, independently of the equipment concerned, either at the very beginning of the project at a time where there are no drawings available or at the qualification stage, in order to define the test conditions.

In the absence of a precise and validated model of the structure, the simplest possible mechanical system is frequently used consisting of mass, stiffness and damping (a linear system with one degree of freedom), especially for:

– the comparison of the severity of several shocks (shock response spectrum) or of several vibrations (extreme response and fatigue damage spectra);

– the drafting of specifications: determining a vibration which produces the same effects on the model as the real environment, with the underlying hypothesis that the equivalent value will remain valid on the real, more complex structure;

– the calculations for pre-sizing at the start of the project;

– the establishment of rules for analysis of the vibrations (choice of the number of calculation points of a power spectral density) or for the definition of the tests (choice of the sweep rate of a swept sine test).

This explains the importance given to this simple model in this work of five volumes on "Vibration and Mechanical Shock":

Volume 1 is devoted to *sinusoidal vibration*. The responses, relative and absolute, of a mechanical one-degree-of-freedom system to an arbitrary excitation are considered, and its transfer function in various forms defined. By placing the properties of sinusoidal vibrations in the contexts of the environment and of laboratory tests, the transitory and steady state response of a single-degree-of-freedom system with viscous and then with non-linear damping is evolved. The various sinusoidal modes of sweeping with their properties are described, and then, starting from the response of a one-degree-of-freedom system, the consequences of an unsuitable choice of the sweep rate are shown and a rule for choice of this rate deduced from it.

Volume 2 deals with *mechanical shock*. This volume presents the shock response spectrum (SRS) with its different definitions, its properties and the precautions to be taken in calculating it. The shock shapes most widely used with the usual test facilities are presented with their characteristics, with indications how to establish test specifications of the same severity as the real, measured environment. A demonstration is then given on how these specifications can be made with classic laboratory equipment: shock machines, electrodynamic exciters driven by a time signal or by a response spectrum, indicating the limits, advantages and disadvantages of each solution.

Volume 3 examines the analysis of *random vibration*, which encompass the vast majority of the vibrations encountered in the real environment. This volume describes the properties of the process enabling simplification of the analysis, before presenting the analysis of the signal in the frequency domain. The definition of the power spectral density is reviewed as well as the precautions to be taken in calculating it, together with the processes used to improve results (windowing, overlapping). A complementary third approach consists of analyzing the statistical properties of the time signal. In particular, this study makes it possible to determine the distribution law of the maxima of a random Gaussian signal and to simplify the calculations of fatigue damage by avoiding direct counting of the peaks (Volumes 4 and 5).

Having established the relationships which provide the response of a linear system with one degree of freedom to a random vibration, Volume 4 is devoted to the calculation of *damage fatigue*. It presents the hypotheses adopted to describe the behaviour of a material subjected to fatigue, the laws of damage accumulation, together with the methods for counting the peaks of the response, used to establish a histogram when it is impossible to use the probability density of the peaks obtained with a Gaussian signal. The expressions of mean damage and of its standard deviation are established. A few cases are then examined using other hypotheses (mean not equal to zero, taking account of the fatigue limit, non linear accumulation law, etc.).

Volume 5 is more especially dedicated to presenting the method of *specification development* according to the principle of tailoring. The extreme response and fatigue damage spectra are defined for each type of stress (sinusoidal vibrations, swept sine, shocks, random vibrations, etc.). The process for establishing a specification as from the life cycle profile of the equipment is then detailed, taking account of an uncertainty factor, designed to cover the uncertainties related to the dispersion of the real environment and of the mechanical strength, and of another coefficient, the test factor, which takes into account the number of tests performed to demonstrate the resistance of the equipment.

This work is intended first and foremost for engineers and technicians working in design teams, which are responsible for sizing equipment, for project teams given the task of writing the various sizing and testing specifications (validation, qualification, certification, etc.) and for laboratories in charge of defining the tests and their performance, following the choice of the most suitable simulation means.

Introduction

Sinusoidal vibrations were the first to be used in laboratory tests to verify the ability of equipment to withstand their future vibratory environment in service without damage. Following the evolution of standards and testing facilities, these vibrations, generally speaking, are currently studied only to simulate vibratory conditions of the same nature, as encountered, for example, in equipment situated close to revolving machinery (motors, transmission shafts, etc). Nevertheless, their value lies in their simplicity, enabling the behaviour of a mechanical system subjected to dynamic stress to be demonstrated and the introduction of basic definitions.

Given that generally speaking the real environment is more or less random in nature, with a continuous frequency spectrum in a relatively wide range, in order to overcome the inadequacies of the initial testing facilities testing rapidly moved to the "swept sine" type. Here the vibration applied is a sinusoid the frequency of which varies over time according to a sinusoidal or exponential law. Despite the relatively rapid evolution of electrodynamic exciters and electrohydraulic vibration exciters, capable of generating wide-band random vibrations, these swept sine standards have lasted, and are in fact still used, for example in aerospace applications. They are also widely used for measuring the dynamic characteristics of structures.

Following a few brief reminders of basic mechanics (Chapter 1), Chapter 2 of this volume on "Sinusoidal Vibration" examines the relative and absolute response of a mechanical system with one degree of freedom subjected to a given excitation, and defines the transfer function in different forms. Chapter 3 is devoted more particularly to the response of such a system to a unit impulse or to a unit step.

The properties of sinusoidal vibrations are then presented in the context of the environment and in laboratory tests (Chapter 4). The transitory and steady state response of a system with one degree of freedom to viscous damping (Chapter 5) and to non-linear damping (Chapter 6) is then examined.

Chapter 7 defines the various sinusoidal sweeping modes, with their properties and eventual justification. Chapter 7 is also devoted to the response of a system with one degree of freedom subjected to linear and exponential sweeping vibrations, to illustrate the consequences of an unsuitable choice of sweep rate, resulting in the presentation of a rule for the choice of rate.

In the appendix are reviewed the major properties of the Laplace transform, which provides a powerful tool for the analytical calculation of the response of a system with one degree of freedom to a given excitation. Inverse transforms more particularly suitable for this application are given in a table.

List of symbols

The list below gives the most frequent definition of the main symbols used in this book. Some of the symbols can have another meaning locally which will be defined in the text to avoid confusion.

$A(t)$	Indicial admittance or step response		F_c	Peak factor (or crest factor)
$A(p)$	Laplace transform of $A(t)$		F_f	Form factor
c	Viscous damping constant		F_m	Maximum value of $F(t)$
c_{eq}	Equivalent viscous damping constant		g	Acceleration due to gravity
			G	Coulomb modulus
$C(\theta)$	Part of the response relating to non-zero initial conditions		$G(\eta)$	Attenuation related to sweep rate
			h	Interval (f/f_0)
D	Damping capacity		H_{AD}	Transmissibility
e	Neper number		H_{RD}	Dynamic amplification factor
E	Young's modulus			
E_d	Dynamic modulus of elasticity		H_{RV}	Relative transmissibility
$E(\ )$	Function characteristic of sweep mode		$h(t)$	Impulse response
			$H(\ )$	Transfer function
f	Frequency of excitation		i	$\sqrt{-1}$
f_m	Expected frequency		I	Moment of inertia
$\dot{f}$	Sweep rate		J	Damping constant
f_0	Natural frequency		k	Stiffness or uncertainty coefficient
$F(t)$	External force applied to a system		ℓ_{rms}	rms value of $\ell(t)$
F_b	Damping force		ℓ_m	Maximum value of $\ell(t)$

$\ell(t)$	Generalized excitation (displacement)
$\dot{\ell}(t)$	First derivative of $\ell(t)$
$\ddot{\ell}(t)$	Second derivative of $\ell(t)$
$L(p)$	Laplace transform of $\ell(t)$
$L(\Omega)$	Fourier transform of $\ell(t)$
m	Mass
n	Number of cycles
n_d	Number of decades
N	Normal force
N_s	Number of cycles performed during swept sine test
p	Laplace variable
P	Reduced pseudo-pulsation
q_m	Maximum value of $q(\theta)$
q_0	Value of $q(\theta)$ for $\theta = 0$
$q(\theta)$	Reduced response
$\dot{q}_0$	Value of $\dot{q}(\theta)$ for $\theta = 0$
$\dot{q}(\theta)$	First derivative of $q(\theta)$
$\ddot{q}(\theta)$	Second derivative of $q(\theta)$
Q	Q factor (quality factor)
$Q(p)$	Laplace transform of $q(\theta)$
R_m	Ultimate tensile strength
R_{om}	Number of octaves per minute
R_{os}	Number of octaves per second
t	Time
t_s	Sweep duration
T	Duration of application of vibration
T_0	Natural period
T_1	Time-constant of logarithmic swept sine
$u(t)$	Generalized response
U_s	Maximum elastic strain energy stored during one cycle
U_{ts}	Elastic strain energy per unit volume
$U(p)$	Laplace transform of $u(t)$
$U(\Omega)$	Fourier transform of $u(t)$
x_m	Maximum value of $x(t)$
$x(t)$	Absolute displacement of the base of a one-degree-of-freedom system
$\dot{x}(t)$	Absolute velocity of the base of a one-degree-of-freedom system
$\ddot{x}_m$	Maximum value of $\ddot{x}(t)$
$\ddot{x}(t)$	Absolute acceleration of the base of a one-degree-of-freedom system
$\ddot{X}(\Omega)$	Fourier transform of $\ddot{x}(t)$
$y(t)$	Absolute displacement response of the mass of a one-degree-of-freedom system
$\dot{y}(t)$	Absolute velocity response of the mass of a one-degree-of-freedom system
$\ddot{y}(t)$	Absolute acceleration response of the mass of a one-degree-of-freedom system
z_m	Maximum value of $z(t)$
z_s	Maximum static relative displacement
$z(t)$	Relative displacement response of the mass of a one-degree-of-freedom system with respect to its base
$\dot{z}(t)$	Relative velocity response
$\ddot{z}(t)$	Relative acceleration response
$Z(p)$	Generalized impedance

δ	Logarithmic decrement		φ	Phase
$\delta_s(\ )$	Dirac delta function		$\lambda(\theta)$	Reduced excitation
Δ	Energy dissipated per unit time		$\Lambda(p)$	Laplace transform of $\lambda(\theta)$
ΔE_d	Energy dissipated by damping in one cycle		μ	Coefficient of friction
			π	3.141 592 65 ...
Δf	Interval of frequency between half-power points		θ	Reduced time $\dot{q}(\theta)$
			θ_b	Reduced sweep rate
ΔN	Number of cycles between half-power points		Θ	Reduced pseudo-period
			σ	Stress
ϵ	Relative deformation		σ_m	Mean stress
$\dot{\epsilon}$	Velocity of relative deformation		ω_0	Natural pulsation ($2 \pi f_0$)
η	Coefficient of dissipation (or of loss) or reduced sweep rate		Ω	Pulsation of excitation ($2 \pi f$)
			ξ	Damping factor
			ξ_{eq}	Equivalent viscous damping factor
			ψ	Phase

Chapter 1

Basic mechanics

1.1. Static effects/dynamic effects

In order to evaluate the mechanical characteristics of materials, it is important to be aware of the nature of stresses [HAU 65]. There are two main load types that need to be considered when doing this:

– those which can be considered as applied statically;

– those which are the subject of a dynamic analysis of signal versus time.

Materials exhibit different behaviours under static and dynamic loads. Dynamic loads can be evaluated according to the following two criteria:

– the load varies quickly if it communicates notable velocities to the particles of the body in deformation, so that the total kinetic energy of the moving masses constitutes a large part of the total work of the external forces; this first criterion is that used during the analysis of the oscillations of the elastic bodies;

– the speed of variation of the load can be related to the velocity of evolution of the plastic deformation process occurring at a time of fast deformation whilst studying the mechanical properties of the material.

According to this last criterion, plastic deformation does not have time to be completed when the loading is fast. The material becomes more fragile as the deformation velocity grows; elongation at rupture decreases and the ultimate load increases (Figure 1.1).

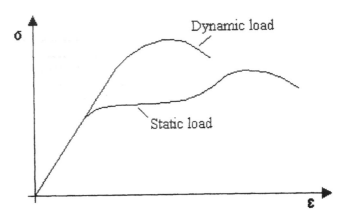

Figure 1.1. *Tension diagram for static and dynamic loads*

Thus, a material can sometimes sustain an important dynamic load without damage, whereas the same load, statically, would lead to plastic deformation or to failure. Many materials subjected to short duration loads have ultimate strengths higher than those observed when they are static [BLA 56], [CLA 49], [CLA 54] and [TAY 46]. The important parameter is in fact the relative deformation velocity, defined by:

$$\dot{\varepsilon} = \frac{1}{\ell_0} \frac{\Delta \ell}{\Delta t} \qquad\qquad [1.1]$$

where $\Delta \ell$ is the deformation observed in time Δt on a test-bar of initial length ℓ_0 subjected to stress.

If a test-bar of initial length 10 cm undergoes in 1 s a lengthening of 0.5 cm, the relative deformation velocity is equal to 0.05 s^{-1}. The observed phenomena vary according to the value of this parameter. Table 1.1 shows the principal fields of study and the usable test facilities. This book will focus on the values in the region 10^{-1} to 10^{1} s^{-1} (these ranges being very approximate).

Certain dynamic characteristics require the data of the dynamic loads to be specified (the order of application being particularly important). Dynamic fatigue strength at any time t depends, for example, on the properties inherent in the material concerning the characteristics of the load over time, and on the previous use of the part (which can reflect a certain combination of residual stresses or corrosion).

Table 1.1. *Fields of deformation velocity*

	0	10^{-5}	10^{-1}	10^{1}	10^{5}
Phenomenon	Evolution of creep velocity in time	Constant velocity of deformation	Response of structure, resonance	Propagation of elastoplastic waves	Propagation of shock waves
Type of test	Creep	Statics	Slow dynamics	Fast dynamics (impact)	Very fast dynamics (hypervelocity)
Test facility	Machines with constant load	Hydraulic machine	Hydraulic vibration machine Shakers	Impact metal - metal Pyrotechnic shocks	Explosives Gas guns
	Negligible inertia forces		Important inertia forces		

1.2. Behaviour under dynamic load (impact)

Hopkinson [HOP 04] noted that copper and steel wire can withstand stresses that are higher than their static elastic limit and are well beyond the static ultimate limit without separating proportionality between the stresses and the strains. This is provided that the length of time during which the stress exceeds the yield stress is of the order of 10^{-3} seconds or less.

From tests carried out on steel (annealed steel with a low percentage of carbon) it was noted that the initiation of plastic deformation requires a definite time when stresses greater than the yield stress are applied [CLA 49]. It was observed that this time can vary between 5 ms (under a stress of approximately 352 MPa) and 6 s (with approximately 255 MPa; static yield stress being equal to 214 MPa). Other tests carried out on five other materials showed that this delay exists only for materials for which the curve of static stress deformation presents a definite yield stress, and the plastic deformation then occurs for the load period.

Under dynamic loading an elastic strain is propagated in a material with a velocity corresponding to the sound velocity c_0 in this material [CLA 54]. This velocity is a function of the modulus of elasticity, E, and of the density, ρ, of material. For a long, narrow part, we have:

$$c_0 = \sqrt{\frac{E}{\rho}}$$

[1.2]

The longitudinal deflection produced in the part is given by:

$$\varepsilon = \frac{v_1}{c_0} \qquad\qquad [1.3]$$

where v_1 = velocity of the particles of material. In the case of plastic deformation, we have [KAR 50]:

$$c(\varepsilon) = \sqrt{\frac{\partial\sigma/\partial\varepsilon}{\rho}} \qquad\qquad [1.4]$$

where $\dfrac{\partial\sigma}{\partial\varepsilon}$ is the slope of the stress deformation curve for a given value of the deformation ε. The velocity of propagation c is, therefore, a function of ε. The relation between the impact velocity and the maximum deformation produced is given by:

$$v_1 = \int_0^{\varepsilon_1} c \, d\varepsilon \qquad\qquad [1.5]$$

A velocity of impact v_1 produces a maximum deformation ε_1 that is propagated with low velocity since the deformation is small. This property makes it possible to determine the distribution of the deformations in a metal bar at a given time.

Most of the materials presents a total ultimate elongation which is larger at impact than for static loading (Figure 1.2).

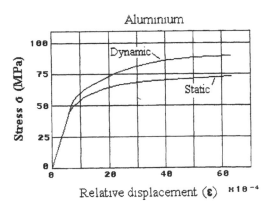

Figure 1.2. *Example of a stress–strain diagram [CAM 53]*

Under the impact, there is deformation of the part and formation of a neck that transmits a very small deformation. This results in a reduction in the total elongation and the energy absorption below the values obtained with low impact velocities, whereas the rupture is not completely brittle. The velocity at which this behaviour occurs is called the *critical impact velocity*.

Some examples of critical velocities are given in Table 1.2.

Table 1.2. *Properties of static and dynamic traction [KAR 50]*

Material	Ultimate strength (10^7 Pa)		Deformation (%)		Critical velocity theoretical (m/s)
	Statics	Dynamics	Statics	Dynamics *	
SAE 5150 hardened and annealing	95.836	102.111	8.5	13.3	48.46
302 standard stainless steel	64.328	76.393	58.5	46.6	149.35
Annealing copper	20.615	25.304	32.7	43.8	70.41
2 S annealing aluminium	7.998	10.618	23.0	30.0	53.65
24S.T aluminium alloy	44.919	47.298	11.3	13.5	88.39
Magnesium alloy (Dow J)	30.164	35.411	9.6	10.9	92.35

* Maximum elongation (in %) to the critical velocity.

1.2.1. *Tension*

The critical impact velocity is specific to the metal under consideration. It is clear from relation [1.5] that there is a maximum impact velocity corresponding to a deformation ε_m for which the slope of the relation $\sigma(\varepsilon)$ is zero (i.e. for the deformation corresponding to the ultimate strength). The velocity of propagation is then zero:

$$(c = \sqrt{\frac{\partial\sigma/\partial\varepsilon}{\rho}}).$$

For the impact velocities that are higher than those corresponding to ε_m, the plastic deformation of the part cannot be propagated as quickly as the end of the part. This results in an immediate rupture between it and the part and no energy is absorbed.

1.3. Elements of a mechanical system

In this section, we will consider lumped parameter systems, in which each particular component can be identified according to its properties and can be distinguished from other elements (in distinction from distributed systems).

Three fundamental passive elements can be defined, each playing a role in linear mechanical systems which corresponds to the coefficients of the expressions of the three types of forces which are opposed to the movement (these parameters can be identified for systems with rotary or translatory movements). These passive elements are frequently used in the modelling of structures to represent a physical system in simple terms [LEV 76].

1.3.1. *Mass*

A *mass* is a rigid body whose acceleration $\ddot{x}$ is, according to Newton's law, proportional to the resultant F of all the forces acting on this body [CRE 65]:

$$F = m \, \ddot{x} \qquad [1.6]$$

It is a characteristic of the body.

In the case of rotational movement, the displacement has the dimension of an angle and acceleration can be expressed in rad/s^2. The constant of proportionality is then called the *moment of inertia* of the body and not mass, although it obeys the same definition. The moment of inertia has the dimension M L. The inertia moment Γ is such that:

$$\Gamma = I_\theta \frac{d^2\theta}{dt^2} \qquad [1.7]$$

where I_θ is the moment of inertia and θ the angular displacement. If $\Omega = \dfrac{d\theta}{dt}$ is the angular velocity we have:

$$\Gamma = I_\theta \frac{d\Omega}{dt} \qquad [1.8]$$

In the SI system, which will be used throughout the book, mass is expressed in kilograms (kg), acceleration in m/s^2 and force in Newtons (N) (dimension MLT^{-2}).

The mass is schematically represented by a rectangle [CHE 66] (Figure 1.3).

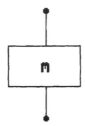

Figure 1.3. *Symbol used to represent mass*

1.3.2. *Stiffness*

1.3.2.1. *Definition*

In the case of linear movement, the *stiffness* of a spring is the ratio k of the variation of force ΔF to the spring deflection Δz which it produces: $k = -\dfrac{\Delta F}{\Delta z}$. The minus sign indicates that the force is opposed to the displacement (*restoring* force) (cf. Figure 1.4).

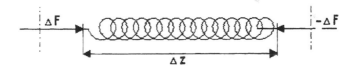

Figure 1.4. *Symbol used for spring*

This definition implicitly assumes that the spring obeys Hooke's law when the deformation is weak.

Stiffness k, *the spring constant,* is evaluated in the SI system n Newtons per metre. It is assumed that the stiffness is that of a perfectly elastic massless spring [CRE 65] [CHE 66]. It is represented diagrammatically by the symbol ꝫꝫꝫꝫꝫꝫ or sometimes ─ＷＷＷＷＷ─ . The points to zero imposed displacement are shown as ⁄⁄⁄⁄⁄⁄⁄⁄⁄ (ground).

In the case of rotation around an axis, the restoring moment is defined by:

$$\Gamma = -C \, \alpha \tag{1.9}$$

with the same convention used for the negative sign. The constant C, which characterizes elasticity here, is expressed in Newtons per radian.

The stiffness of a perfectly rigid medium would thus be theoretically infinite. The input and the output would be identical (the input being, for example, a force transmitted by the medium). The elongation would, of course, be zero. This is a theoretical case, since no material is perfectly rigid. When the stiffness decreases, the *response* of the spring (value function of time obtained at the end of the spring when an input *excitation* is applied at the other end) changes and can become different from the input.

1.3.2.2. *Equivalent spring constant*

Certain systems comprising several elastic elements can be reduced to the simple case of only one spring whose equivalent stiffness can be easily calculated from the general expression $F = -k \, z$. If the system can be moved in just one direction, it can be seen that the number of degrees of freedom is equal to the number of elements of mass. The number of elements of elasticity does not intervene and it is possible to *reduce* the system to a simple unit *mass–spring equivalent*; see Figure 1.5 for some examples.

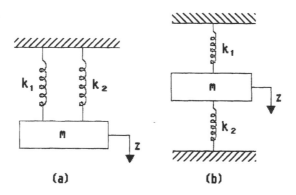

Figure 1.5. *Parallel stiffnesses*

The two diagrams in Figure 1.5 are equivalent to those in Figure 1.6. When the mass is moved by a quantity z, the restoring force is written [CLO 80], [HAB 68] and [VER 67]:

$$|F_r| = k_1 z + k_2 z = k_{eq} z \qquad\qquad [1.10]$$

$$k_{eq} = k_1 + k_2 \qquad\qquad [1.11]$$

The stiffness elements have parallel configuration here.

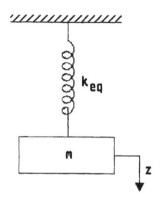

Figure 1.6. *Equivalent stiffness*

In *series* (Figure 1.7), the equivalent constant is calculated in a similar way.

F is a force directed downwards and produces an elongation of each spring respectively equal to:

$$z_1 = \frac{F}{k_1} \qquad\qquad [1.12]$$

and

$$z_2 = \frac{F}{k_2} \qquad\qquad [1.13]$$

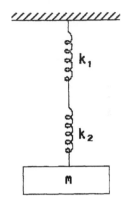

Figure 1.7. *Stiffnesses in series*

This yields

$$k_{eq} = \frac{F}{z} = \frac{F}{z_1 + z_2} = \frac{F}{\dfrac{F}{k_1} + \dfrac{F}{k_2}}$$ [1.14]

i.e

$$\boxed{\frac{1}{k_{eq}} = \frac{1}{k_1} + \frac{1}{k_2}}$$ [1.15]

The equivalent stiffness of two springs in parallel is equal to the sum of their stiffnesses. The inverse of the stiffness of two springs in series is equal to the sum of the inverses of their stiffness [HAB 68] [KLE 71a].

It is easy to generalize this to n springs.

1.3.2.3. *Stiffness of various parts*

Table 1.3 (a). *Examples of stiffnesses [DEN 56], [THO 65a]*

Springs in compression or axial tension		
D = average diameter of a coil d = diameter of the wire n = number of active coils G = elasticity modulus to shearing		$k = \dfrac{G\,d^4}{8\,n\,D^3}$ Deformation: $\delta = \dfrac{8\,F_y\,D^3\,n}{G\,d^4}$
Beam cantilever axial load S = area of the cross-section E = Young modulus		$k = \dfrac{E\,S}{\ell} = \dfrac{F}{X}$
Cantilever beam I = moment of inertia of the section		$k = \dfrac{6\,E\,I}{\ell_0^3\left(3\,\ell - \ell_0\right)} = \dfrac{F}{X}$
Cantilever beam		$k = \dfrac{6\,E\,I}{\ell_0^3\left(3\,\ell - \ell_0\right)} = \dfrac{F}{X}$
Cantilever beam		$k = \dfrac{2\,E\,I}{\ell^2} = \dfrac{M}{X}$
Beam on two simple supports, charged at an arbitrary point		$k = \dfrac{3\,E\,I\,\ell}{\ell_1^2\,\ell_2^2}$

Table 1.3 (b). *Examples of stiffnesses [DEN 56], [THO 65a]*

Fixed beam, loaded in its center		$k = \dfrac{192\ E\ I}{\ell^3}$
Circular plate thickness t, centrally loaded and circumferential edge simply supported		$k = \dfrac{16\ \pi\ D}{R^2}\dfrac{1+\nu}{3+\nu}$ where $D = \dfrac{E\ t^3}{12\left(1-\nu^2\right)}$ ν = Poisson coefficient $(\approx 0.3\ )$
Circular plate centrally loaded, circumferential edge clamped		$k = \dfrac{16\ \pi\ D}{R^2}$

Stiffnesses in rotation

Table 1.4 (a). *Examples of stiffnesses (in rotation)*

Twist of coil spring D = average coil diameter d = wire diameter n = number of turns		$k = \dfrac{E\ d^4}{64\ n\ D}$
Bending of coil spring		$k = \dfrac{E\ d^4}{32\ n\ D}\dfrac{1}{1+E/2\ G}$
Spiral spring ℓ = total length I = moment of inertia of cross section		$k = \dfrac{E\ I}{\ell}$

Table 1.4 (b). *Examples of stiffness (in rotation)*

Twist of hollow circular tube D = outer diameter d = inner diameter ℓ = length		$k = \dfrac{G\,I}{\ell} = \dfrac{\pi\,G}{32}\,\dfrac{D^4 - d^4}{\ell}$ Steel: $k = 1.18\ 10^6\,\dfrac{D^4 - d^4}{\ell}$
Cantilever beam End moment		$k = \dfrac{M}{\theta} = \dfrac{E\,I}{\ell}$
Cantilever beam **End load**		$k = \dfrac{M}{\theta} = \dfrac{2\,E\,I}{\ell^2}$
Beam on two simple supports **Couple at its center**		$k = \dfrac{M}{\theta} = \dfrac{12\,E\,I}{\ell}$
Clamped-clamped **Couple at center**		$k = \dfrac{M}{\theta} = \dfrac{16\,E\,I}{\ell}$
Circular bar D = diameter ℓ = length		$k = \dfrac{\pi\,G\,D^4}{32\,\ell}$
Rectangular plate		$k = \dfrac{G\,w\,t^3}{3\,\ell}$
Bar of arbitrary form S = section I_p = polar inertia moment of the cross-section		$k = \dfrac{G\,S^4}{4\,\pi^2\,\ell\,I_p}$

1.3.2.4. *Non-linear stiffness*

A linear stiffness results in a linear force–deflection curve (Figure 1.8) [LEV 76]. Figures 1.9 to 1.11 show examples of non-linear stiffnesses.

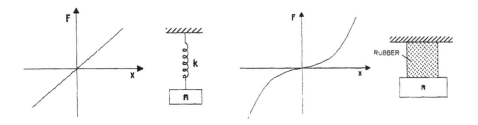

Figure 1.8. *Linear stiffness*   **Figure 1.9.** *Non-linear stiffness*

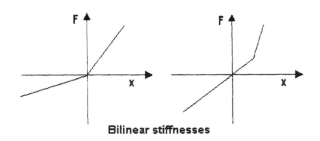

Bilinear stiffnesses

Figure 1.10. *Examples of bilinear stiffnesses*

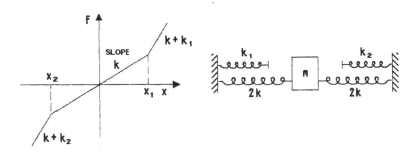

Figure 1.11. *An example of non-linear stiffness*

1.3.3. *Damping*

1.3.3.1. *Definition*

No system subjected to a dynamic stress maintains constant amplitude without an external input of energy. Materials do not behave in a perfectly elastic manner. even at low levels of stress. When cycles of alternate stress (stress varying between a positive maximum and a negative minimum) are out carried on a metal test-bar, we can distinguish the following [BAS 75]:

1. *The microelastic ultimate stress* σ_{me}, such that $\sigma \leq \sigma_{me}$, the stress–strain curve is perfectly linear (zero surface). The stress σ_{me} is, in general, very small.

2. *The anelastic stress* σ_{an} such that $\sigma_{me} < \sigma < \sigma_{an}$, the stress–strain cycle remains closed (without its surface being zero). In this case, the deformation remains 'reversible', but is associated with a dissipation of energy.

3. *The accommodation ultimate stress* σ_{ac}, which is the strongest stress. although the first cycle is not closed. The repetition of several alternate stress cycles can still lead to the closing of the cycle ('accommodation' phenomenon).

4. For $\sigma > \sigma_{ac}$, the cycle is closed, leading to permanent deformation.

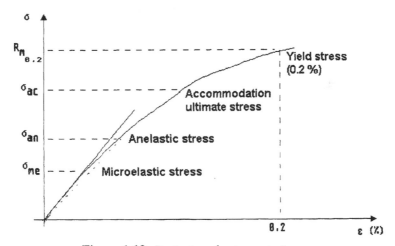

Figure 1.12. *Beginning of a stress–strain curve*

Figure 1.12 shows the beginning of a stress–strain curve. The *yield stress* $R_{m0.2}$ which is defined in general as the stress that produces a deflection of 0.2%, is a conventional limit already located in the plastic range.

There is always a certain inelasticity that exists, although it is often very low and negligible. Due to this inelasticity, which can have several origins, the material or the structure dissipates part of the energy which it receives when a mechanical stress is applied to it. This is said to be *damping*.

Dissipated energy leads to a decrease in the amplitude of the free vibration of the system in the course of time, until it returns to its equilibrium position. This loss is generally connected to the relative movement between components of the system [HAB 68]. The energy communicated to the cushioning device is converted into heat. A damping device is thus a non-conservative device.

The inelastic behaviour is underlined by plotting the stress–strain curve of a test-bar of material subjected to sinusoidal stress (for example in tension–compression) [LAZ 50], [LAZ 68].

Figure 1.13 shows such a curve (very deformed so as to show the phenomenon more clearly).

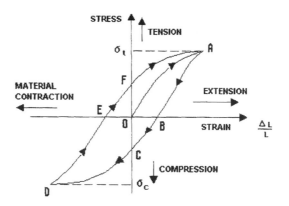

Figure 1.13. *Hysteresis loop*

At the time of the first tension loading, the stress–strain law is represented by arc OA. The passage of tension to compression is done along arc ABCD, while the return towards maximum tension follows arc DEFA.

Curve ABCDEFA is called a *hysteresis loop* and it occurs for completely alternating loads.

1.3.3.2. *Hysteresis*

A natural phenomenon observed in materials, hysteresis is related to partial relaxation of stress applied by means of plastic deformations and characterized by the absorption and the dissipation of energy [FEL 59]. This property of material, studied since being highlighted by Lord Kelvin [THO 65b], has been given various names [FEL 59]:

– *damping capacity* [FÖP 36], term most usually used, which can be defined as the aptitude of material to dissipate vibratory energy; this parameter, denoted by D, defined in 1923 by O. Föppl represents the work dissipated in heat per unit of volume of material during a complete cycle of alternated load and is calculated by evaluating the area delimited by the hysteresis loop [FEL 59]:

$$D = \int_{1 \text{ cycle}} \sigma \, d\varepsilon \qquad\qquad [1.16]$$

thus D is *the absorbed energy by a macroscopically uniform material, per unit of volume and stress cycle* (tension–compression, for example);

– internal friction [ZEN 40], relating to the capacity of a solid to convert its mechanical energy into internal energy;

– mechanical hysteresis [STA 53];

– elastic hysteresis [HOP 12].

Whether for a part comprised of a single material, which may or may not be part of a structure, or for a more complex structure, the hysteresis loop can be plotted by considering the variations of the deformation z consecutive to the application of a sinusoidal force F. The energy dissipated by cycle is then equal to:

$$\Delta E_d = \int_{1 \text{ cycle}} F \, dz \qquad\qquad [1.17]$$

ΔE_d is *the total damping energy* (equal to $\dot{V} D$, where V is the volume of the part). ΔE_d is usually expressed in the following units:

– for a material: Joules per m^3 and cycle;

– for a structure: Joules per cycle.

The total plastic deformation can be permanent anelastic or a combination of both. Hysteresis thus appears by the non-coincidence of the diagram stress–strain loading and unloading curves.

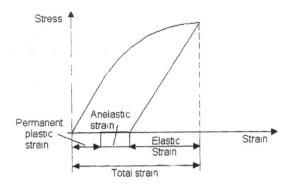

Figure 1.14. *Strain or hysteresis*

If the stress is sufficient to produce plastic deformation, the part will never return to its initial state ($\varepsilon = 0$, $\sigma = 0$). Even if the deformation is only anelastic, there is still the formation of a hysteresis loop. However, if the stress is maintained at zero for a sufficient period of time, the part can return to zero initial condition.

The anelastic strain is, therefore, not just a function of stress, but also of time (as well as temperature, magnetic field, and so on).

1.3.3.3. *Origins of damping*

Damping in materials has been studied for around 200 years. The principal motivations for this has been the analysis of the mechanisms which lead to inelastic behaviour and the dissipation of energy, control with the manufacture of certain characteristics of the materials (purity, fissures, etc.) and especially design of structures, where it is interesting to attenuate the dynamic response stresses.

The damping of a complex structure is dependent on [HAY 72], [LAZ 68]:

– the *internal damping* of the materials which constitute each part;

– the damping introduced by connections or contacts between the various parts (*structural damping*).

Internal damping indicates the complex physical effects which transform into heat the deformation energy in a vibrating mechanical system composed of a volume of macroscopically continuous matter [GOO 76].

When a perfectly elastic system is deformed by the application of an external force, the energy spent by the force during the deformation is stored in the material. When the external force is removed, the stored energy is released and the material oscillates around its equilibrium position (the system is not damped).

In a perfectly plastic material, all the energy spent by the external force is dissipated and no energy is stored in the material. The suppression of the external force thus leaves the material in its deformed state (completely damped system).

Typical materials are neither perfectly elastic, nor perfectly plastic, but partly both. The ratio of the plasticity and the elasticity of a particular material, used to describe the behaviour of this material, is the *damping* or *loss coefficient* of material.

The origins of internal damping are multiple [CRA 62]: dislocations, phenomena related to the temperature, diffusion, magnetomechanical phenomena etc. Damping depends on the level of stress to which the material is subjected, the distribution of the stresses in the specimen, sometimes of frequency, the static load, the temperature, etc. The external magnetic field can also be an important factor for ferromagnetic materials [BIR 77], [FOP 36], [LAZ 68] and [MAC 58]. The effects of these various parameters vary according to whether the inelasticity belongs to the one of the following categories:

1. Inelasticity function of the rate of setting in stress ($\frac{d\sigma}{dt}$ or $\frac{d\varepsilon}{dt}$),

2. Inelasticity independent of the rate of setting in stress,

3. Reversible strain under stress (Figures 1.15-a and 1.15-b),

4. Irreversible strain under stress.

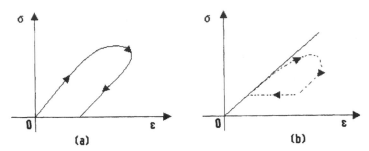

Figure 1.15. *Reversible strain under stress*

These four cases can combine in pairs according to [LAZ 68]:

– case (1 and 3): the material undergoes a strain known as *anelastic*. The anelasticity is characterized by:

    – linear behaviour: to double the stress results in doubling the strain;

    – the existence of a single stress–strain relation, on the condition of allowing sufficient time to reach equilibrium;

– case (1 and 4): the material in this case is known as *viscoelastic*. The viscoelastic strain can be reversible or not. Case (1 and 3) is a particular case of (1 and 4) (recoverable strain);

– case (2 and 4): the material works in a field of plastic strain (under strong stresses in general).

The energy dissipated in cases 1 and 2 can be a function of the amplitude of the stress, but only (2 and 4) is independent of the stress frequency (i.e. of the rate of setting in stress).

Cases (1 and 3) and (1 and 4), for which damping is a function of the loading rate, thus lead to equations which involve the first derivatives $\dfrac{d\sigma}{dt}$ or $\dfrac{d\varepsilon}{dt}$. These cases of damping can be encountered in metals (anelasticity), in the polymers (elastomers) (molecular interaction phenomena), in the structures with various denominations: dynamic hysteresis; rheological damping; and internal friction [LAZ 68].

The relation between applied stress and damping is often complex; one can, however, in a great number of cases, approach this satisfactorily satisfactorily by a relation of the form [LAZ 50], [LAZ 53] and [LAZ 68]:

$$D = J \, \sigma^n \qquad\qquad\qquad\qquad [1.18]$$

where:

J and n are constants for the material;

J = damping constant (or dissipated energy at an unity amplitude stress);

n = damping exponent whose value varies (from 2 to 8) according to the behaviour of the material, related to the stress amplitude, according to the temperature.

The exponent n is a constant for many materials when the stress amplitude is below a certain critical stress which is close to the stress of the ultimate resistance of material.

Above this limit, damping becomes a function of the stress according to time [CRA 62].

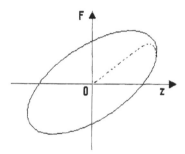

Figure 1.16. *Elliptical cycle (*n = 2*)*

For low stress amplitudes and ambient temperature, n is equal to 2 (*quadratic damping*) and its hysteresis loop has an elliptical form.

One defines the case n = 2 as that of *a linear* damping, because it is observed in the case of a viscosity phenomenon, for which the differential equations describing the movement are linear.

At the intermediate and high amplitudes, one observes non-linear behaviours characterized by non-elliptic hysteresis loops and exponents n generally greater than 2 (observed up to 30 on a material with high stress).

The damping capacity D is defined for a material under uniform stress. The relation [1.18] is in general valid up to a limit called the *limit of sensitivity to the cyclic stresses*, which is in the zone of the fatigue limit of the material [MOR 63a].

Structural damping, the least well-known, is the dominating phenomenon [NEL 80]. The phenomenology of the dissipation of the energy at the ideal simple junction is reasonably well understood [BEA 82], [UNG 73], in particular at low frequencies. The problem is more difficult at high frequencies (much higher than the fundamental resonance frequency of the component). One can schematically distinguish three principal types of interfaces:

– *interfaces with dry friction:* metal–metal or more generally material–material (Coulomb friction); the frictional force, is directly proportional to the normal force and the friction coefficient μ and independent of the sliding velocity. Dissipated energy is equal to the work spent against the frictional force;

– *lubricated interfaces* (fluid film, plastic, etc.) [POT 48]. In this mechanism, the friction is known as *viscous*. The amplitude of the damping force is directly proportional to the velocity of the relative movement and its direction is opposed to that of the displacement;

– interfaces that are *bolted, welded, stuck, rivetted,* etc.

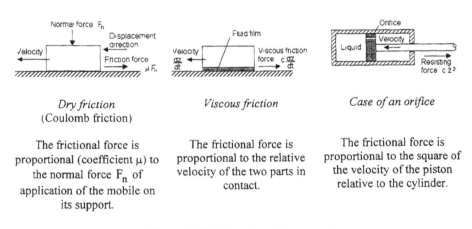

Dry friction
(Coulomb friction)

Viscous friction

Case of an orifice

The frictional force is proportional (coefficient μ) to the normal force F_n of application of the mobile on its support.

The frictional force is proportional to the relative velocity of the two parts in contact.

The frictional force is proportional to the square of the velocity of the piston relative to the cylinder.

Figure 1.17. *Examples of damping forces*

In the first two categories, the forces can be applied in the direction normal to the plane of interface or according to a direction located in the plane of the interface (shearing). It is in this last case that the energy dissipation can be strongest.

There are many other mechanisms of energy dissipation, such as, for example:

– damping due to the environment (air), the moving part activating the air or the ambient fluid (damping force F_d is in general proportional to $\dot{z}^2$);

– magnetic damping (passage of a conductor in a magnetic field; the damping force is then proportional to the velocity of the conductor);

– the passage of a fluid through an orifice, etc.

1.3.3.4. *Specific damping energy*

The *specific damping energy* [FEL 59] is the ratio;

$$\phi = \frac{\Delta E_d}{U_s} \qquad\qquad [1.19]$$

where:

ΔE_d = damping capacity (area under the hysteresis loop);

$$U_s = \frac{\sigma^2}{2\,E_d} = \text{maximum strain energy in the specimen during the cycle;}$$

E_d = dynamic modulus of elasticity.

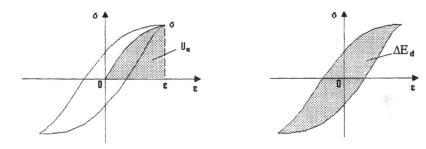

Figure 1.18. *Strain energy*   **Figure 1.19.** *Damping capacity*

The damping of a material can also be defined as the ratio of the dissipated energy to the total strain energy (by cycle and unit of volume):

$$\eta = \frac{D}{2\,\pi\,U_{ts}} \qquad\qquad [1.20]$$

For a linear material, $D = J \sigma^2$ and $U_{ts} = \dfrac{U_s}{V} = \dfrac{1}{2} \dfrac{\sigma^2}{E_d} \left(= \dfrac{1}{2} \dfrac{F}{S} \dfrac{\Delta\ell}{\ell} \right)$,

yielding:

$$\eta = \frac{J E_d}{\pi} \qquad\qquad\qquad [1.21]$$

Constants similar to those used above for viscoelasticity can be defined for anelastic materials. For anelastic materials, η lies between 0.001 and 0.1, while for viscoelastic materials, η varies between 0.1 and 1.5.

1.3.3.5. Viscous damping constant

The theory of viscous internal friction is very old and was employed for a long time. Proposed by Coulomb, it was developed by W.Voight and E.J.Rought, and then used by other authors. It supposes that, in the solid bodies, there are certain viscous attributes which can be compared with the viscosity of a fluid and which are proportional to the first derivative of the deformation [VOL 65]. This yields the damping force:

$$F_d = -c \frac{dz}{dt} \qquad\qquad\qquad [1.22]$$

The factor c, which we will suppose to be constant at first approximation, can vary in practice more or less according to the material, and with the frequency of the excitation. This parameter is a function of the geometry of the damping device and the viscosity of the liquid used. It is encountered at the time of the slip between lubricated surfaces in damping devices with fluid or certain types of laminar flows through an orifice. Damping can be considered to be viscous as long as the flow velocity is not too large.

It is estimated in general that the elastomers and the air cushions (with low velocities) have behaviour comparable to viscous damping. This type of damping is very often used in studies of behaviour of the structures under vibration [JON 70], [JON 69], because it leads to linear equations which are relatively easy to treat analytically.

Viscous damping will be represented on the diagrams by the symbol ⊣⊢ [JON 69].

In the case of a linear system in rotation, the damping torque Γ_d is:

$$\Gamma_d = D_\alpha \ \Omega = D_\alpha \ \frac{d\alpha}{dt} \qquad\qquad [1.23]$$

where:

D_α = viscous damping constant in rotation;

$\Omega = \dfrac{d\alpha}{dt}$ = angular velocity.

1.3.3.6. *Rheology*

Rheology relates to the study of the flow and deformation of matter [ENC 73]. Theoretical rheology attempts to define mathematical models accounting for the behaviour of solids under stresses. The simplest models are those with only one parameter:

– elastic solid of Hooke, with force varying linearly with the displacement, without damping (cf. Figure 1.20);

Figure 1.20. *Elastic solid*

– damping *shock absorber* type of device, with force linearly proportional to the velocity (cf. Figure 1.21).

Among the models, which account better for the behaviour of the real solids, are those models with two parameters [BER 73] such as:

– the Maxwell model, adapted rather well to represent the behaviour of the viscoelastic liquids (cf. Figure 1.22);

Figure 1.21. *Shock absorber-type damping device*

Figure 1.22. *Maxwell model*

– the Kelvin–Voigt model, better adapted to the case of viscoelastic solids. It allows a complex representation of the stiffness and damping for a sinewave excitation of the form:

$$k^* = k + i \Omega c \qquad [1.24]$$

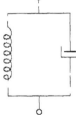

Figure 1.23. *Kelvin–Voigt model*

1.3.3.7. *Damper combinations*

Dampers in parallel

The force necessary to produce a displacement z between the ends of the dampers is equal to:

$$F = F_1 + F_2 = c_1 \, z + c_2 \, z \qquad [1.25]$$

$$F = (c_1 + c_2) \, z = c_{eq} \, z \qquad [1.26]$$

$$\boxed{c_{eq} = c_1 + c_2} \qquad [1.27]$$

Figure 1.24. *Dampers in parallel*

Dampers in series [CLO 80] [VER 67]

$$F = c_1\, z_1 + c_2\, z_2 \qquad\qquad [1.28]$$

$$z = z_1 + z_2 = \frac{F}{c_1} + \frac{F}{c_2} = \frac{F}{c} \qquad\qquad [1.29]$$

$$\boxed{c_{eq} = \frac{1}{1/c_1 + 1/c_2}} \qquad\qquad [1.30]$$

Figure 1.25. *Dampers in series*

1.3.3.8. *Non-linear damping*

Types of non-linear damping are described in Chapter 4 and their effect on the response of a one degree-of-freedom mechanical system is examined. As an example, the case of dry friction (Coulomb damping) and that of an elastoplastic strain [LEV 76] are described.

Dry friction (or Coulomb friction)

The damping force is here proportional to the normal force N between the two moving parts (Figure 1.26):

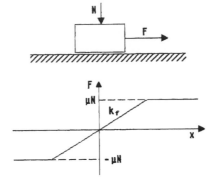

Figure 1.26. *Dry friction*

$$F = \mu \, N \qquad\qquad [1.31]$$

If: $k_f \, x > \mu \, N$

$$F = k_f \, x \qquad\qquad [1.32]$$

If: $-\mu \, N < k_f \, x < \mu \, N$

$$F = -\mu \, N \qquad\qquad [1.33]$$

If: $k_f \, s < -\mu \, N$.

This case will be detailed in Chapter 6.

Elements with plastic deformation

Figures 1.27 and 1.28 show two examples of force–displacement curves where plastic behaviour intervenes.

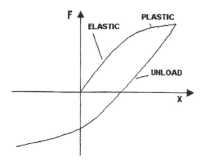

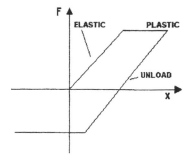

Figure 1.27. *Example of plastic deformation*

Figure 1.28. *Example of plastic deformation*

1.3.4. Static modulus of elasticity

The static modulus of elasticity of a material which is dependent on the stiffness of the parts under the static load in which they are cut, is defined as the ratio of the variation of stress $\Delta\sigma$ to the resulting strain $\varepsilon = \dfrac{\Delta\ell}{\ell}$.

The linear materials have a single module even with very strong damping. For the phenomena which are independent of the rate of setting in stress, such as those observed for metals working under the usual conditions of stress and temperature, the hysteresis loop no longer has an elliptic form which makes it possible to separate the elastic strain component which stores the energy and the component of energy dissipation. Two types of static modules are then defined [LAZ 50], [LAZ 68]:

– *the tangent modulus of elasticity,* for a given value of the stress; this module is proportional to the slope of the stress–strain curve measured for this given stress;

– *the secant modulus of elasticity,* this is proportional to the slope of a straight line segment joining two given points of the stress–strain curve.

As an example, the tangent modulus at the origin is given (Figure 1.29) by the slope of tangent OG to arc OA at the origin and the secant modulus by the slope of segment OA (for a viscous linear material having thus an elliptic hysteresis loop, the secant modulus is none other than the static modulus of elasticity). The tangent OG corresponds to a material which would be perfectly elastic.

In practice, materials in the stress domain have similar tangents and secant moduli of elasticity where they follow Hooke's law reasonably well. In general the secant modulus decreases when the maximum stress amplitude grows.

1.3.5. *Dynamic modulus of elasticity*

The *dynamic modulus of elasticity* of a material is the modulus of elasticity calculated from a stress–strain diagram plotted under cyclic dynamic stress. A tangent dynamic modulus and a secant dynamic modulus are defined in the same way. The values measured in dynamics often differ from static values.

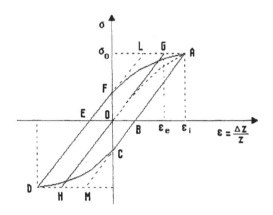

Figure 1.29. *Tangent modulus*

The stress–strain curve can be modified in dynamics by:

– a change in the initial tangent modulus (at the origin) (modification of the slope of OG or any other arc of curve and possibly even by rotation of the hysteresis loop);

– a variation in the surface delimited by the curve, i.e. of the damping capacity of material.

A perfectly elastic material which has a stress–strain curve such as HOG undergoes a strain ε_e under the maximum stress σ_0, whereas the inelastic material takes a deformation ε_i under the same stress. The difference $\Delta\varepsilon = \varepsilon_i - \varepsilon_e$ is a measurement of the dynamic elasticity reduction (to which an increase in the damping capacity corresponds).

When the damping capacity of a material grows, the material becomes more deformed (for the same stress) and its dynamic modulus of elasticity decreases.

These variations can be represented by writing the dynamic module (*secant module*) in the form:

$$E_d = \frac{\sigma_0}{\varepsilon_1} \qquad\qquad [1.34]$$

$$E_d = \frac{\sigma_0}{\varepsilon_e + \Delta\varepsilon} = \frac{1}{\dfrac{\varepsilon_e}{\sigma_0} + \dfrac{\Delta\varepsilon}{\sigma_0}} \qquad\qquad [1.35]$$

$$E_d = \left(\frac{1}{E_e} + \frac{\Delta\varepsilon}{\sigma_0} \right)^{-1} \qquad\qquad [1.36]$$

The initial tangent dynamic module E_e is supposed to be equal to the static module. Since the specific damping capacity D is equal to the area under the curve of the hysteresis loop, we can set:

$$D = K \,\Delta\varepsilon \,\sigma_0 \qquad\qquad [1.37]$$

where K is a constant function of the shape of the cycle (for example, K = 4 for a trapezoidal cycle such as LAMDL). This yields:

$$E_d = \left(\frac{1}{E_e} + \frac{D}{K \,\sigma_0^2} \right)^{-1} \qquad\qquad [1.38]$$

The value of K depends on the shape of the loop, as well as on the stress amplitude. An average value is K = 3 [LAZ 50].

The modulus E_d is thus calculable from [1.38] provided that the initial tangent modulus (or the slope of arcs DF or AC) does not vary (with the velocity of loading, according to the number of cycles, etc.). B.J.Lazan [LAZ 50] has shown that in a particular case this variation is weak and that expression [1.38] is sufficiently precise.

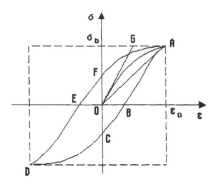

Figure 1.30. *Tangent and secant moduli*

– slope of OG = *initial tangent modulus of elasticity* (i.e. approximately the slope of arcs AB and OF which are appreciably linear);

– slope of OA = *secant modulus of elasticity*;

– OB = *remanent deformation*;

– OF = *coercive force.*

1.4. Mathematical models

1.4.1. *Mechanical systems*

A *system* is a unit made up of mechanical elements having properties of mass, stiffness and of damping. The mass, stiffness and damping of a structure are important parameters since they determine its dynamic behaviour.

The system can be:

– a *lumped parameter system* when the components can be isolated by distinguishing the masses, stiffnesses and dampings, by supposing them lumped in separate elements. In this case, the position at a given time depends on a finite number of parameters;

– a *distributed system,* when this number is infinite. The movement is then a function of time and space.

1.4.2. *Lumped parameter systems*

In practice, and generally for a real structure, these elements are distributed continuously, uniformly or not, with the properties of mass, stiffness and damping not being separate. The structure is made up of an infinite number of infinitesimal particles. The behaviour of such a system with distributed constants must be studied using complete differential equations with partial derivatives.

It is often interesting to simplify the structure to be studied in order to be able to describe its movement using complete ordinary differential equations, by dividing it into a discrete number of specific masses connected by elastic massless elements and of energy dissipative elements, so as to obtain a lumped parameter system [HAB 68], [HAL 78].

The transformation of a physical system with distributed constants into a model with localized constants is in general a delicate operation, with the choice of the points having an important effect on the results of the calculations carried out thereafter with the model.

The procedure consists of:

1. Choosing a certain number of points (nodes) by which the mass of the structure is affected. The number of nodes and number of directions in which each node can be driven determines the number of degrees of freedom of the model.

The determination of the number of nodes and their position can be a function of:

– the nature of the study to be carried out: to roughly define a problem, it is often enough to be limited to a model with a few degrees of freedom;

– the complexity of the structure studied;

– available calculation means: if the complexity of the structure and the precision of the results justify it, then a model with several hundred nodes can be considered.

The choice of the number of nodes is therefore, in general, a compromise between a sufficient representativeness of the model and a simple analysis, leading to the shortest possible computing time.

2. Distributing the total mass of the structure between the various selected points. This task must be carried out carefully, particularly when the number of nodes is limited.

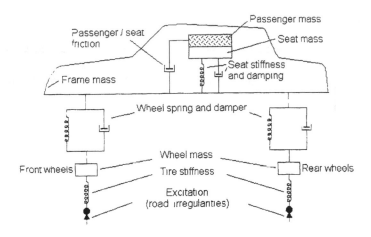

Figure 1.31. *Mathematical model of a car*

This type of modelling makes it easier to study more complicated structures such as a car-passenger unit (Figure 1.31) [CRE 65]. Such a model is sometimes called a *mathematical model.*

In these models, according to the preceding definitions, the element mass is supposed to be perfect, i.e. perfectly rigid and non-dissipative of energy, the elasticity element is massless and perfectly elastic, finally the dissipative energy element is supposed to be perfectly massless and rigid.

Computer programmes have been developed to study numerically the dynamic behaviour of structures modelled in this way[GAB 69] [MAB 84] [MUR 64].

1.4.3. *Degrees of freedom*

The number of degrees of freedom of a material system is equal to the number of parameters necessary at any time to determine the state of this system. The simplest system, a material point, has three degrees of freedom in general: three co-ordinates are necessary at every moment to define its position in space. The number of equations necessary to know the movement of the system must be equal to the number of degrees of freedom (dof).

A solid has six degrees of freedom in general. This number depends on:

– the complexity of the solid;

– the connections to which it is subjugated.

If each element of mass of a model can be driven only in only one direction, the number of degrees of freedom is equal to the number of elements of mass. A very complex system can thus have a limited number of degrees of freedom.

(NOTE: *A deformable system has an infinite number of degrees of freedom.*)

1.4.4. *Mode*

The exploitation of these models with lumped or distributed constants show that the system can vibrate in a certain number of ways, called *modes*. Each one corresponds to a specific natural frequency. This number of frequencies is therefore equal to the number of modal shapes, and is therefore equal to the number of co-ordinates necessary at any moment to determine the position of the system, i.e., according to paragraph 1.4.3, to the number of degrees of freedom of the system.

In the case of a system with distributed masses, the number of degrees of freedom is infinite. Each frequency corresponds to a single oscillatory mode, which is determined by its characteristic function or normal function. A transient or permanent forced excitation will excite, in general, some or all of these frequencies, the response in each point being a combination of the corresponding modal shapes. In the case of a linear system, one will be able to use the principle of superposition to calculate this response.

This concept of mode is important and deserves further development. The chapters which follow are limited to systems with only one degree of freedom.

Examples

1. Case of a beam fixed at an end, length L, uniform section and bending stiffness EI (E = modulus of elasticity and I = inertia moment of the section).

The natural pulsation ω_0 is given by [CRE 65]:

$$\omega_0 = n^2 \pi^2 \sqrt{\frac{g\,E\,I}{P\,L^4}} = n^2 \pi^2 \sqrt{\frac{E\,I}{m\,L^4}} \qquad [1.39]$$

where n is an integer: n = 1, 2, 3...
   g is the acceleration of gravity (9.81 m/s^2)

yielding frequencies

$$f_0 = \frac{n^2 \pi}{2} \sqrt{\frac{g E I}{P L^4}} = K \sqrt{\frac{g E I}{P L^4}} \quad \text{(Hertz)} \qquad [1.40]$$

where P is the weight of the beam per unit of length. Each value of n corresponds a frequency f_0. Figure 1.32 shows the first five modes.

Mode 1	Mode 2	Mode 3	Mode 4	Mode 5
K = 0.56	K = 3.51	K = 9.82	K = 19.24	K = 31.81

Figure 1.32. *First five modes of a fixed beam*

2. Beam fixed at the two ends [STE 78]:

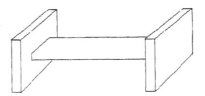

Figure 1.33. *Beam fixed at both ends*

Natural frequency

$$f_0 = \frac{22,44}{2 \pi} \sqrt{\frac{E I g}{P L^3}} \qquad [1.41]$$

where

E = Young's modulus (units SI)   I = moment of inertia
P = weight of the beam   L = length of the beam
g = 9.81 m/s²

Coupled modes

In a system with several degrees of freedom, the mode of one of the degrees can influence the movement corresponding to that of another degree.

It is important to distinguish between coupled and uncoupled movements. When two movements of a mass, horizontally and vertically for example, are not coupled, and can coexist simultaneously and independently, the system is not regarded as one with several degrees of freedom, but composed of several systems with only one degree of freedom whose movements are collectively used to obtain the total resulting movement [CRE 65].

1.4.5. *Linear systems*

A vibrating *linear system* is any system whose positional variables follow in the absence of an external exciting force, a system of linear differential equations, with constant coefficients, and no second members, in a number equal to that of the unknown factors [MAZ 66].

In a linear structure, the characteristics of the response are additive and homogeneous [PIE 64]:

– the response to a sum of excitations is equal to the sum of the responses to each individual excitation;

– the response to k times the excitation (k = constant) is equal to k times the response to the excitation.

This concept of linearity generally imposes an assumption of weak displacements (for example, small relative displacement response of the mass of a one-degree-of-freedom system).

1.4.6. *Linear one-degree-of-freedom mechanical systems*

The simplest mechanical system consists of mass, a stiffness and a damping device (Voigt model) (Figure 1.34). The response is calculated using a linear differential equation of the second order. Due to its simplicity, the results can be expressed in concise form, with a limited number of parameters.

The one-degree-of-freedom system is a model used for the analysis of mechanical shocks and vibrations (comparison of the severity of several excitations of the same nature or different nature, development of specifications, etc.). The implicit idea is that if a vibration (or a shock A) leads to a relative displacement response larger than a vibration B on a one-degree-of-freedom system, vibration A will be more severe than B on a more complex structure.

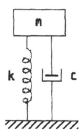

Figure 1.34. *Voigt model*

The displacement of an arbitrary system subjected to a stress being in general primarily produced by the response of the lowest frequency, this one-degree-of-freedom model very often makes it possible to obtain a good approximation to the result. For more precise stress calculations, the use of a more complicated mathematical model is sometimes necessary.

1.4.7. *Setting an equation for n degrees-of-freedom lumped parameter mechanical system*

Various methods can be used to write the differential equations of the movement of a several degrees-of-freedom mechanical system with localized constants. One of these consists of the study of the action and reaction forces on each mass when a mass is moved slightly. To avoid the possible errors of sign during the evaluation of these forces when the system is complex, the following rule can be used [STE 73].

For each mass m_i of the model, it is written that all the forces associated with the mass m_i are positive and that all the forces associated with the other masses m_j ($j \neq i$) are negative.

In practice, for each mass m_i, the sum of the damping, spring and inertia forces is made equal to zero as follows:

– *inertial force*: positive, equal to $m_i \ddot{y}_i$;

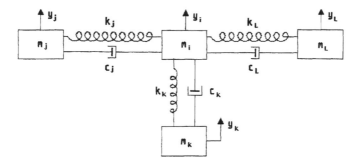

Figure 1.35. *Example of a lumped parameter system*

– *restoring force*: equal to $k_i\left(y_i - y_j\right)$, with k_i being the stiffness of each elastic element connected to the mass m_i, $y_i - y_j$ being written while starting with the co-ordinate y_i of the mass m_i, and y_j being the co-ordinate of the other end of each spring;

– *damping force*: same rule as for the stiffnesses, with the first derivative of $c_i\left(\dot{y}_i - \dot{y}_j\right)$.

The mass m_i (Figure 1.35), is as follows:

$$m_i\,\ddot{y}_i + c_j\left(\dot{y}_1 - \dot{y}_j\right) + k_j\left(y_i - y_j\right) + c_k\left(\dot{y}_i - \dot{y}_k\right) + k_k\left(y_i - y_k\right)$$
$$+c_\ell\left(\dot{y}_i - \dot{y}_\ell\right) + k_\ell\left(y_i - y_\ell\right) = 0$$

Chapter 2

Response of a linear single-degree-of-freedom mechanical system to an arbitrary excitation

2.1. Definitions and notation

Any mechanical system can be represented by a combination of the three pure *elements:* mass, spring and damping device (Chapter 1). This chapter examines the movement of the simplest possible systems comprising one, two or three of these different elements when they are displaced from their rest position at an initial instant of time. The movement of mass alone is commonplace and without practical or theoretical interest for our applications. The cases of a spring alone, a damping device alone or a damper–spring system, are really not of little more practical interest than any real system having just a mass. The simplest systems which are of interest are those composed of:

– mass and spring;

– mass, spring and damping.

We will consider the spring and damping to be linear and that the mass m can move in a frictionless manner along a vertical axis (for example) [AKA 69]. This system can be excited by:

– a force applied to the mass m, with the spring and the damping device being fixed to a rigid support (Figure 2.1-a);

– a movement (displacement, velocity or acceleration) of a massless rigid moving support (Figure 2.1-b).

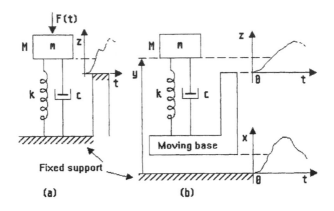

Figure 2.1. *Mass–spring–damping system*

One variable is enough at each instant of time to define the position of the mass m on the axis 0z since it is *a single-degree-of-freedom system*. The origin of the abscissa of the mass m is the point where the mass is at rest (unstretched spring). Gravity is neglected, even if the axis 0z is vertical (as in the case of Figure 2.1). It can be seen that the movement of m around its new static equilibrium position is the same as that around the rest position, excluding gravity.

Note that:

$x(t)$	is the absolute displacement of the support with respect to a fixed reference (Figure 2.1-b).
$\dot{x}(t)$ and $\ddot{x}(t)$	are the corresponding velocity and acceleration.
$y(t)$	is the absolute displacement of the mass m with respect to a fixed reference.
$\dot{y}(t)$ and $\ddot{y}(t)$	are the corresponding velocity and acceleration.
$z(t)$	is the relative displacement of the mass relative to the support. To consider only the variations of z around this position of equilibrium (point 0) and to eliminate length from the spring at rest, the support was drawn so that it goes up to point 0.
$\dot{z}(t)$ and $\ddot{z}(t)$	are the corresponding velocity and acceleration.

F(t)   is the force applied directly to the mass m (Figure 2.1-a).

NOTE: *In the case of Figure 2.1-a, we have y ≡ z.*

The movement considered is a small excitation around the equilibrium position of the system. The excitations $x(t)$, $\dot{x}(t)$, $\ddot{x}(t)$ or $F(t)$, can be of a different nature: i.e. sinusoidal, swept sine, random or shock.

2.2. Excitation defined by force versus time

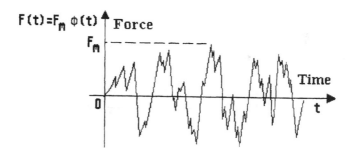

Figure 2.2. *Force versus time*

Let $F(t)$ be the force applied to the mass m of a mass–spring–damping oscillator (single-degree-of-freedom system) [BAR 6] [FUN 58] (Figure 2.3).

We will express the excitation $F(t)$ in a dimensionless form $\lambda(t)$ such that:

$$F(t) = F_m \, \lambda(t)$$

$$\lambda(t) = 0 \quad \text{for } t \le 0$$

$$\max \lambda(t) = \lambda(t_m) = 1$$

[2.1]

The spring is supposed to be linear in the elastic range, with a fixed end and with the other connected to the mass.

The forces which act on the mass m are:

– inertia $m \dfrac{d^2 z}{dt^2}$;

– an elastic force due to the spring, of value $-k\,z$ (restoring force), so long as it follows Hooke's law of proportionality between the forces and deformations. This force is directed in the opposite direction to the displacement;

– a resistant damping force, proportional and opposing the velocity $\dfrac{dz}{dt}$ of the mass, $-c\,\dfrac{dz}{dt}$;

– the imposed external force $F(t)$.

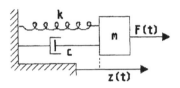

Figure 2.3. *Force on a one-degree-of-freedom system*

The resultant of the forces acting on the mass m (the spring and the damping device of our model being supposed to be ideal and without mass), $-k\,z(t) - c\,\dfrac{dz}{dt} + F(t)$, thus obeys, according to d'Alembert's principle:

$$m \frac{d^2 z}{dt^2} = -k\,z - c\,\frac{dz}{dt} + F(t)$$ [2.2]

This is the *differential equation of the movement of the vibrating one-degree-of-freedom system* [DEN 60], yielding:

$$\frac{d^2 z}{dt^2} + \frac{c}{m}\frac{dz}{dt} + \frac{k}{m}\,z = \frac{F(t)}{m}$$ [2.3]

If we set

$$\xi = \frac{c}{2\,m\,\omega_0} \qquad\qquad [2.4]$$

and

$$\omega_0^2 = \frac{k}{m} \qquad\qquad [2.5]$$

it becomes:

$$\frac{d^2z}{dt^2} + 2\,\xi\,\omega_0\,\frac{dz}{dt} + \omega_0^2\,z = \frac{F_m}{m}\,\lambda(t) \qquad\qquad [2.6]$$

ω_0 is *the natural pulsation* of the system or *angular frequency* and is expressed in radians per second (the pulsation of the undamped oscillator when the mass is moved away from its equilibrium position). It is only a characteristic of the system as long as the *small oscillations* assumption is checked (i.e. as long as it can be supposed that the potential energy is a function of the square of the coordinate) [POT 48].

The natural period of the system is defined as:

$$T_0 = \frac{2\,\pi}{\omega_0} \qquad\qquad [2.7]$$

and *the natural frequency*

$$f_0 = \frac{\omega_0}{2\,\pi} \qquad\qquad [2.8]$$

where T_0 is expressed in seconds (or its submultiples) and f_0 in Hertz (1 Hz = 1 cycle/s).

ξ is the *damping factor* or *damping ratio*: $\xi = \dfrac{c}{2\,m\,\omega_0} = \dfrac{c}{2\,\sqrt{k\,m}}$.

NOTE: *When the mass moves horizontally without friction on a perfectly smooth surface, we do not have to consider other forces. The rest position is then at the same time the equilibrium position of the mass and the unstretched position of the spring.*

If we suppose that the mass is suspended on the spring along to a vertical axis, an application of the equation can be made, either by only considering the equilibrium position, or by considering the rest position of the spring.

If we count the amplitude z starting from the equilibrium position, i.e. starting from position 0, where the force of gravity m g *is balanced by the spring force* k z_{eq} *(*z_{eq} *being the deflection of the spring due to the gravity g, measured from the point 0), its inclusion in the equation is absolutely identical to that of the preceding paragraphs.*

If, on the contrary, we count amplitude z from the end of the spring in its rest position 0, z_1 *is equal to* $z + z_{eq}$ *; it is thus necessary to replace z by* $z + z_{eq}$ *in equation [2.6] and to add in the second member, a force* mg. *After simplification (*k z_{eq} = mg*), the final result is of course the same. In all the following paragraphs, regardless of excitation, the force* mg *will not be taken into account.*

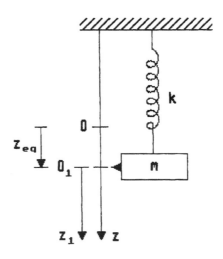

Figure 2.4. *Equilibrium position*

If we set u = z and

$$\ell(t) = \frac{F(t)}{k} = \frac{F_m}{k} \lambda(t)$$

[2.9]

then equation [2.6] can be written in the form:

$$\boxed{\ddot{u} + 2\,\xi\,\omega_0\,\dot{u} + \omega_0^2\,u = \ell(t)\,\omega_0^2}$$

[2.10]

2.3. Excitation defined by acceleration

The support receives *an excitation*, which we will suppose to be defined by a known acceleration $\ddot{x}(t)$. The excitation is propagated towards the mass through the elements k and c. The disturbance which m undergoes is translated by a *response movement*.

Excitation and response are not independent entities, but are mathematically related (Figure 2.5).

We will suppose that:

− the simple single-degree-of-freedom system (Figure 2.6) is such that mass and base are driven in the same direction:

− the movement $x(t)$ of the support is not affected by the movement of the equipment which it supports.

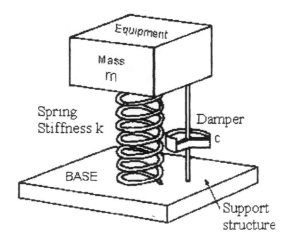

Figure 2.5. *One-degree-of-freedom system*

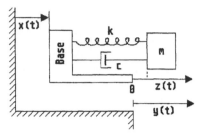

Figure 2.6. *Acceleration on the one-degree-of-freedom system*

The excitation is this known movement $x(t)$ of the support or the acceleration $\ddot{x}(t)$ communicated to this support. The equation for the movement is written:

$$m\frac{d^2y}{dt^2} = -k(y-x) - c\left(\frac{dy}{dt} - \frac{dx}{dt}\right)$$ [2.11]

i.e. using the same notation as before:

$$\frac{d^2y(t)}{dt^2} + 2\,\xi\,\omega_0\,\frac{dy(t)}{dt} + \omega_0^2\,y(t) = \omega_0^2\,x(t) + 2\,\xi\,\omega_0\,\frac{dx(t)}{dt}$$ [2.12]

The relative displacement of the mass relative to the base is equal to

$$z(t) = y(t) - x(t)$$ [2.13]

yielding, after the elimination of y (absolute displacement of m) and its derivative,

$$\boxed{\frac{d^2z}{dt^2} + 2\,\xi\,\omega_0\,\frac{dz}{dt} + \omega_0^2\,z(t) = -\frac{d^2x}{dt^2}}$$ [2.14]

If we set $u = z$ and $\ell(t) = -\dfrac{\ddot{x}}{\omega_0^2}$ (generalized excitation), the equation above can be written as:

[2.15]

$$\boxed{\ddot{u}(t) + 2\,\xi\,\omega_0\,\dot{u} + \omega_0^2\,u = \omega_0^2\,\ell(t)}$$

This similar equation for an excitation by force or acceleration is known as the *generalized form*.

2.4. Reduced form

2.4.1. *Excitation defined by a force on a mass or by an acceleration of support*

Let us set, according to case, $z_s = -\dfrac{\ddot{x}_m}{\omega_0^2}$ or $z_s = \dfrac{F_m}{k}$. This parameter is the maximum static relative displacement, indeed:

$$z_s = \frac{F_m}{k} = \frac{\max \left|F(t)\right|}{k} = \frac{\max\left|F(t)\right|}{m \; \omega_0^2} \tag{2.16}$$

$$z_s = -\frac{\ddot{x}_m}{\omega_0^2} = -\frac{m \; \ddot{x}_m}{m \; \omega_0^2} = \frac{\text{max. of static force corresponding to the max. of } \ddot{x}(t)}{m \; \omega_0^2} \tag{2.17}$$

Let us note that $\ell_m = z_s = \dfrac{\ell(t)}{\lambda(t)}$

$$(\ell_m = \text{maximum of } \ell(t) = \left|\begin{array}{c} -\dfrac{\ddot{x}_m}{\omega_0^2} \\[2mm] \dfrac{F_m}{k} \end{array}\right|),$$

from [2.6] or [2.14]:

$$\frac{\ddot{u}(t)}{\ell_m} + 2 \, \xi \, \omega_0 \; \frac{\dot{u}(t)}{\ell_m} + \omega_0^2 \, \frac{u(t)}{\ell_m} = \omega_0^2 \, \frac{\ell(t)}{\ell_m} \tag{2.18}$$

NOTE: $\ell(t)$ *has the dimension of a displacement.*

Let us set $q = \dfrac{u}{\ell_m}$:

$$\ddot{q}(t) + 2 \, \xi \, \omega_0 \; \dot{q} + \omega_0^2 \, q = \omega_0^2 \, \lambda(t) \tag{2.19}$$

and

$$\omega_0 \, t = \theta \tag{2.20}$$

$$\frac{dq}{dt} = \frac{dq}{d\theta}\frac{d\theta}{dt} = \omega_0 \frac{dq}{d\theta}$$

$$\frac{d^2q}{dt^2} = \frac{d^2q}{d\theta^2}\left(\frac{d\theta}{dt}\right)^2 = \omega_0^2 \frac{d^2q}{d\theta^2}$$

yielding:

$$\boxed{\frac{d^2q}{d\theta^2} + 2\,\xi\,\frac{dq}{d\theta} + q(\theta) = \lambda(\theta)}$$ [2.21]

Table 2.1. *Reduced variables*

System	Excitation		Amplitude of the excitation	Reduced response
	Real	Generalized $\ell(t)$	ℓ_m	$q(t)$
Fixed base	$F(t)$	$\dfrac{F(t)}{k}$	$z_s = \dfrac{F_m}{k}$	$\dfrac{z(t)}{\ell_m}$
Moving base	$\ddot{x}(t)$	$-\dfrac{\ddot{x}(t)}{\omega_0^2}$	$z_s = -\dfrac{\ddot{x}_m}{\omega_0^2}$	$\dfrac{z(t)}{\ell_m}$

A problem of vibration or shock transmitted to the base can thus be replaced by the problem of force applied to the mass of the resonator.

2.4.2. *Excitation defined by velocity or displacement imposed on support*

We showed in [2.12] that the equation of the movement of the system can be put in the form:

$$\ddot{y} + 2\,\xi\,\omega_0\,\dot{y} + \omega_0^2\,y = 2\,\xi\,\omega_0\,\dot{x} + \omega_0^2\,x$$ [2.22]

By double differentiation we obtain:

$$\frac{d^2\ddot{y}}{dt^2} + 2\,\xi\,\omega_0\,\frac{d\ddot{y}}{dt} + \omega_0^2\,\ddot{y} = 2\,\xi\,\omega_0\,\frac{d^2\dot{x}}{dt^2} + \omega_0^2\,\frac{d^2x}{dt^2}$$ [2.23]

If the excitation is a displacement $x(t)$ and if the response is characterized by the absolute displacement $y(t)$ of the mass, the differential equation of the movement [2.22] can be written as [2.24], while setting:

$$\ell(t) = x(t)$$

$$u(t) = y(t)$$

$$\ddot{u} + 2\,\xi\,\omega_0\,\dot{u} + \omega_0^2\,u = 2\,\xi\,\omega_0\,\dot{\ell} + \omega_0^2\,\ell \qquad\qquad [2.24]$$

If the excitation is the velocity $\dot{x}(t)$ and if $\dot{y}(t)$ is the response, the equation [2.23] is written as [2.25], while noting:

$$\ell(t) = \dot{x}(t)$$

$$u(t) = \dot{y}(t)$$

$$\ddot{u} + 2\,\xi\,\omega_0\,\dot{u} + \omega_0^2\,u = 2\,\xi\,\omega_0\,\dot{\ell} + \omega_0^2\,\ell \qquad\qquad [2.25]$$

In the same way, if the input is acceleration $\ddot{x}(t)$ and the response $\ddot{y}(t)$, we have, with:

$$\ell(t) = \ddot{x}(t)$$

$$u(t) = \ddot{y}(t)$$

$$\ddot{u} + 2\,\xi\,\omega_0\,\dot{u} + \omega_0^2\,u = 2\,\xi\,\omega_0\,\dot{\ell} + \omega_0^2\,\ell \qquad\qquad [2.26]$$

This equation is thus another generalized form applicable to a movement imposed on the base and an *absolute response*.

Reduced form

Let us set, as before, $\ell(t) = \ell_m\,\lambda(t)$ [ℓ_m = maximum of $\ell(t)$] and $\omega_0\,t = \theta$, this then becomes:

$$\boxed{\ddot{q}(\theta) + 2\,\xi\,\dot{q}(\theta) + q(\theta) = 2\,\xi\,\dot{\lambda}(\theta) + \lambda(\theta)} \qquad\qquad [2.27]$$

NOTE: *If $\xi = 0$, then equations [2.21] and [2.27] take the single form*

$$\ddot{q}(\theta) + q(\theta) = \lambda(\theta)$$

The excitation for the relative motion is simply the inertial force $m\,\ddot{x}(t)$ required by the adoption of an accelerating frame of reference [CRA 58]. We will find an application of this property in the study of shock response spectra [LAL 75]. These reduced forms could be used for the solution of the equations. The following table indicates the input and response parameters corresponding to the variables $\ell(t)$ and $u(t)$ [SUT 68].

Table 2.2. *Values of the reduced variables*

System	Excitation $\ell(t)$		Response $u(t)$
Fixed base	Force on the	$F(t)/k$	Mass relative displacement $z(t)$
	mass	$F(t)$	Reaction force on base $F_T(t)$
Moving base	Base displacement $\dot{x}(t)$		Mass absolute displacement $y(t)$
	Base velocity $\dot{x}(t)$		Mass absolute velocity $\dot{y}(t)$
	Base acceleration	$\ddot{x}(t)$	Mass absolute acceleration $\ddot{y}(t)$
		$-\dfrac{\ddot{x}(t)}{\omega_0^2}$	Relative displacement of spring $z(t)$
		$m\,\ddot{x}(t)$	Reaction force on base $F_T(t)$

The solution of these two types of differential equation will make it possible to solve all the problems set by these inputs and responses. In practice, however, one will have to choose between the two formulations according to the parameter response desired, which is in general the relative displacement, related to the stress in the simple system.

The more current case is when the excitation is an acceleration. Equation [2.21] is then essential. If the excitation is characterized by a base displacement, the differential equation will in response provide the absolute mass displacement. To return to the stresses we will have to calculate the relative displacement $y - x$.

2.5. Solution of the differential equation of movement

2.5.1. *Methods*

When the excitation can be expressed in a suitable analytical form, the differential equation of the movement can be explicitly solved for $q(\theta)$ or $u(t)$. When it is not the case, the response must be sought using analogue or digital techniques.

The solution $q(\theta)$ can be obtained either by the traditional method of the variation of the constants, or by using the properties of Fourier or Laplace transforms. It is this last, faster method, that we will generally use in the following paragraphs and chapters.

Duhamel integral

A more general method consists of solving the differential equation in the case of an arbitrary excitation $\lambda(\theta)$ with, for example, the Laplace transform. The solution $q(\theta)$ can then be expressed in the form of an integral which, according to the nature of $\lambda(\theta)$ (numerical data, function leading to an analytically integrable expression), can be calculated numerically or analytically.

2.5.2. *Relative response*

2.5.2.1. *General expression for response*

The Laplace transform of the solution of a differential second order equation of the form:

$$\frac{d^2q}{d\theta^2} + a \; \frac{dq}{d\theta} + b \; q(\theta) = \lambda(\theta) \qquad\qquad [2.28]$$

can be written as (appendix) [LAL 75]:

$$Q(p) = \frac{\Lambda(p) + p \; q_0 + a \; q_0 + \dot{q}_0}{p^2 + a \; p + b} \qquad\qquad [2.29]$$

where $\Lambda(p)$ is the Laplace transform of $\lambda(\theta)$

$$\Lambda(p) = L[\lambda(\theta)]$$ [2.30]

$$q_0 = q(0)$$

$$\dot{q}_0 = \dot{q}(0)$$

and, in our case:

$$a = 2\xi$$

$$b = 1$$

After solution of the rational fraction $\dfrac{p\,q(0) + 2\,\xi\,q(0) + \dot{q}(0)}{p^2 + 2\,\xi\,p + 1}$ into simple

elements, [2.29] becomes, with p_1 and p_2 the roots of $p^2 + 2\,\xi\,p + 1 = 0$,

$$Q(p) = \frac{\Lambda(p)}{p_1 - p_2}\left[\frac{1}{p - p_1} - \frac{1}{p - p_2}\right] + \frac{1}{p_1 - p_2}\left[\frac{q_0\,p_1 + 2\,\xi\,q_0 + \dot{q}_0}{p - p_1} - \frac{q_0\,p_2 + 2\,\xi\,q_0 + \dot{q}_0}{p - p_2}\right]$$

[2.31]

The response $q(\theta)$ is obtained by calculating the original of $Q(p)$ [LAL 75]:

$$q(\theta) = \int_0^\theta \frac{\lambda(\alpha)}{p_1 - p_2}\left[e^{p_1\,(\theta - \alpha)} - e^{p_2\,(\theta - \alpha)}\right]d\alpha$$

$$+ \frac{1}{p_1 - p_2}\left[\left(q_0\,p_1 + 2\,\xi\,q_0 + \dot{q}_0\right)e^{p_1\,\theta} - \left(q_0\,p_2 + 2\,\xi\,q_0 + \dot{q}_0\right)e^{p_2\,\theta}\right] \quad [2.32]$$

where α is a variable of integration.

Particular case

For a system at rest at the initial time:

$$q_0 = \dot{q}_0 = 0$$ [2.33]

Then:

$$\boxed{q(\theta) = \int_0^\theta \frac{\lambda(\alpha)}{p_1 - p_2}\left[e^{p_1\,(\theta - \alpha)} - e^{p_2\,(\theta - \alpha)}\right]d\alpha}$$ [2.34]

The movement $q(\theta)$ is different according to nature of the roots p_1 and p_2 of $p^2 + 2\,\xi\,p + 1 = 0$.

2.5.2.2. *Subcritical damping*

In this case, the roots p_1 and p_2 of the denominator $p^2 + 2\xi p + 1$ are complex:

$$p_{1,2} = -\xi \pm i\sqrt{1-\xi^2} \quad \text{(i.e. } 0 \le \xi < 1)$$

[2.35]

While replacing p_1 and p_2 by these expressions, the response $q(\theta)$ given by [2.32] becomes:

$$q(\theta) = \frac{1}{\sqrt{1-\xi^2}} \int_0^\theta \lambda(\alpha) \, e^{-\xi(\theta-\alpha)} \, \sin\sqrt{1-\xi^2}\,(\theta-\alpha)\, d\alpha$$

[2.36]

For a system initially at rest, $q(\theta)$ is reduced to:

$$q(\theta) = \frac{1}{\sqrt{1-\xi^2}} \int_0^\theta \lambda(\alpha) \, e^{-\xi(\theta-\alpha)} \, \sin\sqrt{1-\xi^2}\,(\theta-\alpha)\, d\alpha$$

[2.37]

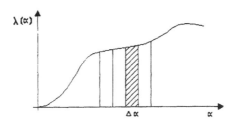

Figure 2.7. *Elemental impulses*

This integral is named *Duhamel's integral* or *superposition integral* or *convolution integral*. We will indeed see that the excitation can be regarded as a series of impulses of duration $\Delta\alpha$ and that the total response can be calculated by superimposing the responses to all these impulses.

For a small damping $\xi \ll 1$

$$q(\theta) \approx \int_0^\theta \lambda(\alpha) \, e^{-\xi(\theta-\alpha)} \, \sin(\theta-\alpha)\, d\alpha$$

[2.38]

and for zero damping:

$$q(\theta) = \int_0^\theta \lambda(\alpha) \sin(\theta - \alpha) \, d\alpha \qquad\qquad [2.39]$$

2.5.2.3. Critical damping

The roots p_1 and p_2 are both equal to -1. This case occurs for $\xi = 1$. The differential equation of movement is then written:

$$\ddot{q}(\theta) + 2\,\dot{q}(\theta) + q(\theta) = \lambda(\theta) \qquad\qquad [2.40]$$

$$L(\ddot{q}) = p^2\,Q(p) - p\,q_0 - \dot{q}_0 \qquad\qquad [2.41]$$

$$L(\dot{q}) = p\,Q(p) - q_0 \qquad\qquad [2.42]$$

$$p^2\,Q(p) - p\,q_0 - \dot{q}_0 + 2\,p\,Q(p) - 2\,q_0 + Q(p) = \Lambda(p) \qquad\qquad [2.43]$$

yielding:

$$Q(p) = \frac{\Lambda(p)}{p^2 + 2p + 1} + \frac{p q_0 + \dot{q}_0 + 2 q_0}{p^2 + 2p + 1} \qquad\qquad [2.44]$$

$$Q(p) = \frac{\Lambda(p)}{(p+1)^2} + \frac{p\,q_0 + \dot{q}_0 + 2\,q_0}{(p+1)^2} = \frac{\Lambda(p)}{(p+1)^2} + \frac{q_0}{p+1} + \frac{\dot{q}_0 + q_0}{(p+1)^2} \qquad [2.45]$$

i.e.

$$\boxed{q(\theta) = \int_0^\theta \lambda(\alpha)\,(\theta - \alpha)\,e^{-(\theta-\alpha)}\,d\alpha + \left[q_0 + \left(q_0 + \dot{q}_0\right)\theta\right]e^{-\theta}} \qquad [2.46]$$

2.5.2.4. Supercritical damping

The roots are real $(\xi > 1)$:

$$q(\theta) = \int_0^\theta \frac{\lambda(\alpha)}{p_1 - p_2}\left[e^{p_1(\theta-\alpha)} - e^{p_2(\theta-\alpha)}\right]d\alpha$$

$$+\frac{1}{p_1 - p_2}\left[\left(p_1\, q_0 + 2\,\xi\, q_0 + \dot{q}_0\right) e^{p_1\,\theta} - \left(p_2\, q_0 + 2\,\xi\, q_0 + \dot{q}_0\right) e^{p_2\,\theta}\right] \quad [2.47]$$

where

$$p_{1,\,2} = \xi \pm \sqrt{\xi^2 - 1} \qquad\qquad\qquad [2.48]$$

since

$$p^2 + 2\,\xi\, p + 1 = p^2 + 2\,\xi\, p + \xi^2 - \xi^2 + 1 \qquad\qquad [2.49]$$

$$p^2 + 2\,\xi\, p + 1 = \left(p + \xi\right)^2 - \left(\xi^2 - 1\right) = \left(p + \xi + \sqrt{\xi^2 - 1}\right)\left(p + \xi - \sqrt{\xi^2 - 1}\right)$$
$$[2.50]$$

$$p_1 - p_2 = 2\sqrt{\xi^2 - 1} \qquad\qquad\qquad [2.51]$$

yielding

$$q(\theta) = \frac{1}{\sqrt{\xi^2 - 1}} \int_0^\theta \lambda(\alpha)\, e^{-\xi\,(\theta - \alpha)}\, \mathrm{sh}\sqrt{\xi^2 - 1}\,(\theta - \alpha)\, d\alpha$$

$$+\frac{e^{-\xi\,\theta}}{\sqrt{\xi^2 - 1}}\left[\left(\xi\, q_0 + \dot{q}_0\right) \mathrm{sh}\sqrt{\xi^2 - 1}\,\theta + q_0\,\sqrt{\xi^2 - 1}\, \mathrm{ch}\sqrt{\xi^2 - 1}\,\theta\right] \quad [2.52]$$

2.5.3. Absolute response

2.5.3.1. General expression for response

The solution of a differential second order equation of the form:

$$\frac{d^2 q}{d\theta^2} + a\,\frac{dq}{d\theta} + b\, q(\theta) = \lambda(\theta) + b\,\frac{d\lambda}{d\theta} \qquad\qquad [2.53]$$

has as a Laplace transform $Q(p)$ [LAL 75]:

$$Q(p) = \frac{\Lambda(p)\left(1 + a\, p\right) + p\, q_0 + a\left(q_0 - \lambda_0\right) + \dot{q}_0}{p^2 + a\, p + b} \qquad [2.54]$$

where

$$q_0 = q(0) \qquad \lambda_0 = \lambda(0) \qquad a = 2\,\xi$$

$$\dot{q}_0 = \dot{q}(0) \qquad \Lambda(p) = L[\lambda(\theta)] \qquad b = 1$$

As previously:

$$Q(p) = \frac{\Lambda(p)}{p^2 + 2\,\xi\,p + 1} + \frac{2\,\xi\,\Lambda(p)}{p^2 + 2\,\xi\,p + 1} \qquad [2.55]$$

$$+\frac{1}{p_1 - p_2}\left[\frac{q_0\,p_1 + 2\,\xi\,(q_0 - \lambda_0) + \dot{q}_0}{p - p_1} - \frac{q_0\,p_2 + 2\,\xi\,(q_0 - \lambda_0) + \dot{q}_0}{p - p_2}\right]$$

$q(\theta)$ is obtained by searching the original of $Q(p)$

$$q(\theta) = \int_0^\theta \frac{\lambda(\alpha)}{p_1 - p_2}\left[(1 + 2\,\xi\,p_1)\,e^{p_1\,(\theta-\alpha)} - (1 + 2\,\xi\,p_2)\,e^{p_2\,(\theta-\alpha)}\right]d\alpha$$

$$+\frac{1}{p_1 - p_2}\left\{\left[q_0\,p_1 + 2\,\xi\,(q_0 - \lambda_0) + \dot{q}_0\right]e^{p_1\,\theta} - \left[q_0\,p_2 + 2\,\xi\,(q_0 - \lambda_0) + \dot{q}_0\right]e^{p_2\,\theta}\right\}$$

$$[2.56]$$

(α = variable of integration).

Particular case

$$\lambda_0 = q_0 = \dot{q}_0$$

$$q(\theta) = \int_0^\theta \frac{\lambda(\alpha)}{p_1 - p_2}\left[(1 + 2\,\xi\,p_1)\,e^{p_1\,(\theta-\alpha)} - (1 + 2\,\xi\,p_2)\,e^{p_2\,(\theta-\alpha)}\right]d\alpha$$

$$[2.57]$$

2.5.3.2. Subcritical damping

The roots of $p^2 + 2\,\xi\,p + 1$ are complex $(0 \le \xi < 1)$

$$p_{1,2} = -\xi \pm i \sqrt{1-\xi^2} \tag{2.58}$$

While replacing p_1 and p_2 by their expressions in $q(\theta)$, it becomes:

$$q(\theta) = \frac{1}{\sqrt{1-\xi^2}} \int_0^\theta \lambda(\alpha)\, e^{-\xi(\theta-\alpha)} \left\{ \left(1-2\,\xi^2\right) \sin\sqrt{1-\xi^2}\,(\theta-\alpha) \right.$$

$$\left. +2\,\xi\,\sqrt{1-\xi^2}\,\cos\sqrt{1-\xi^2}\,(\theta-\alpha) \right\} d\alpha$$

$$+e^{-\xi\theta} \left[q_0 \cos\sqrt{1-\xi^2}\,\theta + \frac{\dot{q}_0 + \xi\left(q_0 - 2\,\lambda_0\right)}{\sqrt{1-\xi^2}} \sin\sqrt{1-\xi^2}\,\theta \right] \tag{2.59}$$

If $\lambda_0 = q_0 = \dot{q}_0 = 0$

$$q(\theta) = \frac{1}{\sqrt{1-\xi^2}} \int_0^\theta \lambda(\alpha)\, e^{-\xi(\theta-\alpha)} \left\{ 2\,\xi\,\cos\sqrt{1-\xi^2}\,(\theta-\alpha) \right.$$

$$\left. +\left(1-2\,\xi^2\right) \sin\sqrt{1-\xi^2}\,(\theta-\alpha) \right\} d\alpha \tag{2.60}$$

If moreover $\xi = 0$

$$q(\theta) = \int_0^\theta \lambda(\alpha) \sin(\theta-\alpha)\, d\alpha \tag{2.61}$$

2.5.3.3. Critical damping

The equation $p^2 + 2\,\xi\,p + 1 = 0$ has a double root $\left(p = -1\right)$. In this case $\xi = 1$ and

$$q(\theta) = \frac{\Lambda(p)\left(1+2\,p\right) + p\,q_0 + \dot{q}_0 + 2\left(q_0 - \lambda_0\right)}{\left(p+1\right)^2} \tag{2.62}$$

$$q(\theta) = \int_0^\theta \lambda(\alpha)\,[2 - \theta + \alpha]\, e^{-(\theta-\alpha)}\, d\alpha + \left[q_0 + \theta\left(q_0 + \dot{q}_0 - 2\,\lambda_0\right)\right] e^{-\theta} \tag{2.63}$$

2.5.3.4. *Supercritical damping*

The equation $p^2 + 2\,\xi\,p + 1 = 0$ has two real roots. This condition is carried out when $\xi > 1$. Let us replace $p_1 = -\xi + \sqrt{\xi^2 - 1}$ and $p_2 = -\xi - \sqrt{\xi^2 - 1}$ by their expressions in [2.56] [KIM 26]:

$$q(\theta) = \int_0^\theta \frac{\lambda(\alpha)}{2\sqrt{\xi^2 - 1}} \left\{ \left[1 + 2\,\xi \left(-\xi + \sqrt{\xi^2 - 1} \right) \right] e^{\left(-\xi + \sqrt{\xi^2 - 1} \right)(\theta - \alpha)} \right.$$

$$\left. - \left[1 + 2\xi \left(-\xi - \sqrt{\xi^2 - 1} \right) \right] e^{\left(-\xi - \sqrt{\xi^2 - 1} \right)(\theta - \alpha)} \right\} d\alpha$$

$$+ \frac{1}{2\sqrt{\xi^2 - 1}} \left\{ \left[\left(-\xi + \sqrt{\xi^2 - 1} \right) q_0 + 2\,\xi\,(q_0 - \lambda) + \dot{q}_0 \right] e^{\left(-\xi + \sqrt{\xi^2 - 1} \right)\theta} \right.$$

$$\left. - \left[\left(-\xi - \sqrt{\xi^2 - 1} \right) q_0 + 2\xi(q_0 - \lambda_0) + \dot{q}_0 \right] e^{\left(-\xi - \sqrt{\xi^2 - 1} \right)\theta} \right\} \qquad [2.64]$$

yielding

$$q(\theta) = \int_0^\theta \frac{\lambda(\theta)}{\sqrt{\xi^2 - 1}} \, e^{-\xi(\theta - \alpha)} \left[\left(1 - 2\xi^2 \right) \mathrm{sh}\sqrt{\xi^2 - 1}\,(\theta - \alpha) \right.$$

$$\left. + 2\,\xi\sqrt{\xi^2 - 1} \; \mathrm{ch}\sqrt{\xi^2 - 1}\,(\theta - \alpha) \right] d\alpha + C(\theta) \qquad [2.65]$$

where

$$C(\theta) = \frac{e^{-\xi\theta}}{\sqrt{\xi^2 - 1}} \left\{ \left[\xi\,(q_0 - 2\,\lambda_0) + \dot{q}_0 \right] \frac{e^{\theta\sqrt{\xi^2 - 1}} - e^{-\theta\sqrt{\xi^2 - 1}}}{2} \right.$$

$$\left. + \sqrt{\xi^2 - 1}\; q_0 \; \frac{e^{\theta\sqrt{\xi^2 - 1}} + e^{-\theta\sqrt{\xi^2 - 1}}}{2} \right\} \qquad [2.66]$$

Another form

$$C(\theta) = \frac{e^{-\xi\,\theta}}{2\sqrt{\xi^2-1}}\left\{e^{\sqrt{\xi^2-1}\,\theta}\left[\xi\left(q_0-2\,\lambda_0\right)+\dot{q}_0+\sqrt{\xi^2-1}\;q_0\right]\right.$$
$$\left.+e^{-\sqrt{\xi^2-1}\,\theta}\left[\sqrt{\xi^2-1}\;q_0-\xi\left(q_0-2\,\lambda_0\right)-\dot{q}_0\right]\right\}$$ [2.67]

$$C(\theta) = a\;e^{\left(-\xi+\sqrt{\xi^2-1}\right)\theta}+b\;e^{\left(-\xi-\sqrt{\xi^2-1}\right)\theta}$$ [2.68]

with

$$a = \frac{\xi\left(q_0-2\,\lambda_0\right)+\dot{q}_0+\sqrt{\xi^2-1}\;q_0}{2\sqrt{\xi^2-1}}$$ [2.69]

$$b = \frac{\sqrt{\xi^2-1}\;q_0-\xi\left(q_0-2\,\lambda_0\right)-\dot{q}_0}{2\sqrt{\xi^2-1}}$$ [2.70]

If $q_0 = \dot{q}_0 = \lambda_0 = 0$:

$$q(\theta) = \int_0^\theta \frac{\lambda(\alpha)}{\sqrt{\xi^2-1}}\;e^{-\xi\,(\theta-\alpha)}\left[\left(1-2\,\xi^2\right)\mathrm{sh}\sqrt{\xi^2-1}\,(\theta-\alpha)\right.$$
$$\left.+2\,\xi\,\sqrt{\xi^2-1}\;\mathrm{ch}\,\sqrt{\xi^2-1}(\theta-\alpha)\right]d\alpha$$ [2.71]

If moreover $\xi = 0$:

$$q(\theta) = \int_0^\theta \lambda(\alpha)\;\mathrm{sh}\,(\theta-\alpha)\;d\alpha$$ [2.72]

2.5.4. *Summary of main results*

Zero initial conditions:

Relative response

$0 \leq \xi < 1$

$$q(\theta) = \frac{1}{\sqrt{1-\xi^2}} \int_0^\theta \lambda(\alpha) e^{-\xi(\theta-\alpha)} \sin \sqrt{1-\xi^2}(\theta-\alpha)\, d\alpha \qquad [2.73]$$

$\xi = 1$ $\qquad\qquad\qquad\qquad\qquad\qquad\qquad\qquad\qquad\qquad\qquad$ [2.74]

$$q(\theta) = \int_0^\theta \lambda(\alpha)(\theta-\alpha) e^{-(\theta-\alpha)}\, d\alpha$$

$\xi > 1$ $\qquad\qquad\qquad\qquad\qquad\qquad\qquad\qquad\qquad\qquad\qquad$ [2.75]

$$q(\theta) = \frac{1}{\sqrt{1-\xi^2}} \int_0^\theta \lambda(\alpha) e^{-\xi(\theta-\alpha)} \operatorname{sh} \sqrt{\xi^2-1}(\theta-\alpha)\, d\alpha$$

Absolute response

$0 \leq \xi < 1$

$$q(\theta) = \frac{1}{\sqrt{1-\xi^2}} \int_0^\theta \lambda(\alpha) e^{-\xi(\theta-\alpha)} \left\{ 2\xi \cos\sqrt{1-\xi^2}(\theta-\alpha) \right.$$

$$\left. + \left(1-2\xi^2\right) \sin\sqrt{1-\xi^2}(\theta-\alpha) \right\} \qquad [2.76]$$

$\xi = 1$

$$q(\theta) = \int_0^\theta \lambda(\alpha)(2-\theta+\alpha) e^{-(\theta-\alpha)}\, d\alpha \qquad [2.77a]$$

$\xi > 1$

$$q(\theta) = \frac{1}{\sqrt{1-\xi^2}} \int_0^\theta \lambda(\alpha) e^{-\xi(\theta-\alpha)} \left[\left(1-2\xi^2\right) \operatorname{sh}\sqrt{\xi^2-1}(\theta-\alpha) \right.$$

$$\left. +2\xi\sqrt{\xi^2-1} \operatorname{ch}\sqrt{\xi^2-1}(\theta-\alpha) \right] d\alpha \qquad [2.77b]$$

If the initial conditions are not zero, we have to add to these expressions. according to the nature of the response.

For $0 \le \xi < 1$

Relative response

$$C(\theta) = e^{-\xi \theta} \left[q_0 \cos \sqrt{1-\xi^2} \theta + \frac{q_0 \xi + \dot{q}_0}{\sqrt{1-\xi^2}} \sin \sqrt{1-\xi^2} \theta \right] \qquad [2.78]$$

Absolute response

$$C(\theta) = e^{-\xi \theta} \left[q_0 \cos\sqrt{1-\xi^2} \theta + \frac{\dot{q}_0 + \xi (q_0 - 2\lambda_0)}{\sqrt{1-\xi^2}} \sin \sqrt{1-\xi^2} \theta \right] \qquad [2.79]$$

For $\xi = 1$

Relative response $\quad C(\theta) = \left[q_0 + \theta (q_0 + \dot{q}_0) \right] e^{-\theta} \qquad [2.80]$

Absolute response $\quad C(\theta) = \left[q_0 + (q_0 + \dot{q}_0 - 2\lambda_0) \theta \right] e^{-\theta} \qquad [2.81]$

For $\xi > 1$

Relative response

$$C(\theta) = e^{-\xi \theta} \left[\frac{(\xi q_0 + \dot{q}_0)}{\sqrt{\xi^2 - 1}} \text{sh} \sqrt{\xi^2 - 1} \theta + q_0 \text{ ch} \sqrt{\xi^2 - 1} \theta \right] \qquad [2.82]$$

Absolute response

$$C(\theta) = e^{-\xi \theta} \left[\frac{(\xi q_0 + \dot{q}_0 - 2 \xi \lambda_0)}{\sqrt{\xi^2 - 1}} \text{sh} \sqrt{\xi^2 - 1} \theta + q_0 \text{ ch} \sqrt{\xi^2 - 1} \theta \right] \quad [2.83]$$

In all these relations, the only difference between the cases $0 \le \xi < 1$ and $\xi > 1$ resides in the nature of the sine and cosine functions (hyperbolic for $\xi > 1$).

2.6. Natural oscillations of a linear single-degree-of-freedom system

We have just shown that the response $q(\theta)$ can be written for non-zero initial conditions:

$$q_{IC}(\theta) = q(\theta) + C(\theta) \qquad [2.84]$$

The response $q_{IC}(\theta)$ is equal to the sum of the response $q(\theta)$ obtained for zero initial conditions and the term $C(\theta)$ corresponds to a damped oscillatory response produced by non-zero initial conditions. So $\lambda(\theta) = 0$ is set in the differential equation of the movement [2.21]:

$$\ddot{q}(\theta) + 2\,\xi\,\dot{q}(\theta) + q(\theta) = 0$$

it then becomes, after a Laplace transformation,

$$Q(p) = \frac{\dot{q}_0 + 2\,\xi\,q_0 + p\,q_0}{p^2 + 2\,\xi\,p + 1} \qquad (0 \le \xi < 1) \qquad [2.85]$$

The various cases related to the nature of the roots of:

$$p^2 + 2\,\xi\,p + 1 = 0 \qquad [2.86]$$

are considered here.

2.6.1. Damped aperiodic mode

In this case $\xi > 1$. The two roots of $p^2 + 2\,\xi\,p + 1 = 0$ are real. Suppose that the response is defined by an absolute movement. If this were not the case, it would be enough to make $\lambda_0 = 0$ in the relations of this paragraph. The response $q(\theta)$ of the system around its equilibrium position is written:

$$q(\theta) = e^{-\xi\,\theta} \left[\frac{\dot{q}_0 + \xi\,(q_0 - 2\,\lambda_0)}{\sqrt{\xi^2 - 1}} \; \text{sh} \; \sqrt{\xi^2 - 1}\;\theta + q_0 \; \text{ch} \; \sqrt{\xi^2 - 1}\;\theta \right] \qquad [2.87]$$

$q(\theta)$ can also be written in the form:

$$q(\theta) = a \; e^{\left(-\xi+\sqrt{\xi^2-1}\right)\theta} + b \; e^{-\left(\xi+\sqrt{\xi^2-1}\right)\theta} \qquad [2.88]$$

where

$$a = \frac{\xi\,(q_0 - 2\,\lambda_0) + \dot{q}_0 + \sqrt{\xi^2 - 1}\;q_0}{2\,\sqrt{\xi^2 - 1}} \qquad [2.89]$$

and

$$b = \frac{q_0\,\sqrt{\xi^2 - 1} - \xi\,(q_0 - 2\,\lambda_0) - \dot{q}_0}{2\,\sqrt{\xi^2 - 1}} \qquad [2.90]$$

It is noticed that the roots $-\xi + \sqrt{\xi^2 - 1}$ and $-\xi - \sqrt{\xi^2 - 1}$ of the equation $p^2 + 2\xi p + 1 = 0$ are both negative, their sum negative and their product positive. So the two exponential terms are decreasing functions of time, like $q(\theta)$.

The velocity $\dfrac{dq}{d\theta}$, which is equal to:

$$\frac{dq}{d\theta} = a\left(-\xi + \sqrt{\xi^2 - 1}\right) e^{\left(-\xi + \sqrt{\xi^2 - 1}\right)\theta} - b\left(\xi + \sqrt{\xi^2 - 1}\right) e^{-\left(\xi + \sqrt{\xi^2 - 1}\right)\theta} \qquad [2.91]$$

is also, for the same reason, a decreasing function of time. Therefore, the movement cannot be oscillatory. It is *a damped exponential motion*. $q(\theta)$ can also be written:

$$q(\theta) = a\, e^{\left(-\xi + \sqrt{\xi^2 - 1}\right)\theta} \left[1 + \frac{b}{a} e^{\left(-\xi - \sqrt{\xi^2 - 1} + \xi - \sqrt{\xi^2 - 1}\right)\theta}\right] \qquad [2.92]$$

$$q(\theta) = a\, e^{\left(-\xi + \sqrt{\xi^2 - 1}\right)\theta} \left[1 + \frac{b}{a} e^{-2\sqrt{\xi^2 - 1}\,\theta}\right] \qquad [2.93]$$

When θ tends towards infinity, $e^{-2\sqrt{\xi^2 - 1}}$ tends towards zero $(\xi > 1)$. After a certain time, the second term thus becomes negligible in comparison with the unit and $q(\theta)$ behaves then like:

$$a\, e^{\left(-\xi + \sqrt{\xi^2 - 1}\right)\theta} \qquad [2.94]$$

As $\left(-\xi + \sqrt{\xi^2 - 1}\right)$ is always negative, $q(\theta)$ decreases constantly according to time.

If the system is moved away from its equilibrium position and released with a zero velocity $\dot{q}_0$ with an elongation q_0 at time $t = 0$, coefficient a becomes, for $\lambda_0 = 0$

$$a = q_0 \frac{\xi + \sqrt{\xi^2 - 1}}{2\sqrt{\xi^2 - 1}} \qquad [2.95]$$

q_0 being supposed positive, a being positive and $q(\theta)$ always remaining positive: the system returns towards its equilibrium position without crossing it.

The velocity can also be written:

$$\frac{dq}{d\theta} = a\left(-\xi + \sqrt{\xi^2 - 1}\right) e^{\left(-\xi + \sqrt{\xi^2 - 1}\right)\theta} \left[1 - \frac{b}{a} \frac{\xi + \sqrt{\xi^2 - 1}}{-\xi + \sqrt{\xi^2 - 1}} e^{-2\sqrt{\xi^2 - 1}\,\theta}\right]$$

[2.96]

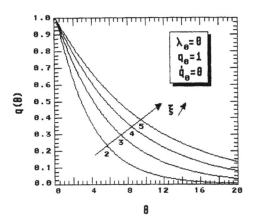

Figure 2.8. *Damped aperiodic response*

We have:

$$\frac{dq}{d\theta} \approx a\left(-\xi + \sqrt{\xi^2 - 1}\right) e^{\left(-\xi + \sqrt{\xi^2 - 1}\right)\theta}$$

[2.97]

when θ is sufficiently large. The velocity is then always negative.

Variations of the roots p_1 and p_2 according to ξ

The characteristic equation [2.86] $p^2 + 2\xi p + 1 = 0$ is that of a hyperbole (in the axes p, ξ) whose asymptotic directions are:

$$p^2 + 2\xi p = 0$$

i.e.

$$\begin{cases} p = 0 \\ p + 2\,\xi = 0 \end{cases}$$ [2.98]

The tangent parallel with the axis 0p is given by $2\,p + 2\,\xi = 0$, i.e. $p = -\xi$.

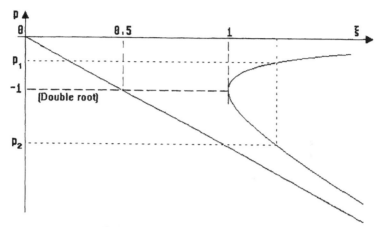

Figure 2.9. *Characteristic equation*

This yields $\xi = 1$ while using [2.86] (since ξ is positive or zero), i.e.

$$c = c_c = 2\,\sqrt{k\,m}$$ [2.99]

Any parallel with the axis 0p such as $\xi > 1$ crosses the curve at two points corresponding to the values p_1 and p_2 of p.

The system returns all the more quickly to its equilibrium position, so that $q(\theta)$ decreases quickly, therefore $|p_1|$ is larger (the time-constant, in the expression $q(\theta) = a\,e^{\left(-\xi + \sqrt{\xi^2 - 1}\right)\theta}$, is of value $\left|\dfrac{1}{p_1}\right| = \dfrac{1}{\left|-\xi + \sqrt{\xi^2 - 1}\right|}$), i.e. the relative damping ξ is still smaller (or the coefficient of energy dissipation c).

$|p_1|$ has the greatest possible value when the equation $p^2 + 2\,\xi\,p + 1 = 0$ has a double root, i.e. when $\xi = 1$.

NOTE: *If the system is released out of its equilibrium position with a zero initial velocity, the resulting movement is characterized by a velocity which changes only once in sign. The system tends to its equilibrium position without reaching it. It is said that the motion is 'damped aperiodic'* $(\xi > 1)$. *Damping is supercritical.*

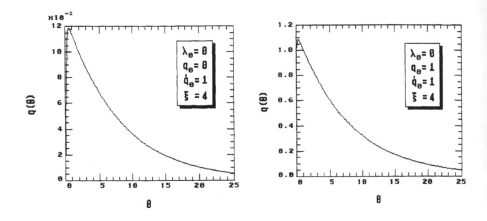

Figure 2.10. *Aperiodic damped response (for $q_0 = 0$)*   **Figure 2.11.** *Aperiodic damped response (for $q_0 = 1$)*

2.6.2. Critical aperiodic mode

On the assumption that $\xi = 1$, the two roots of $p^2 + 2\xi p + 1 = 0$ are equal to -1. By definition

$$\xi = \frac{c}{2\sqrt{k\,m}} = 1 \qquad\qquad [2.100]$$

yielding $c = c_c = 2\sqrt{km}$. The parameter c_c, the *critical damping coefficient*, is the smallest value of c for which the damped movement is non-oscillatory. It is the reason why ξ is also defined as the *fraction of critical damping* or *critical damping ratio*.

According to whether the response is relative or absolute, the response $q(\theta)$ is equal to:

$$q(\theta) = \left[q_0 + \left(q_0 + \dot{q}_0 \right) \theta \right] e^{-\theta} \qquad\qquad [2.101]$$

or

$$q(\theta) = \left[q_0 + \left(q_0 + \dot{q}_0 - 2\,\lambda_0 \right) \theta \right] e^{-\theta} \qquad\qquad [2.102]$$

As an example, Figures 2.12–2.14 show $q(\theta)$, respectively, for $\lambda_0 = 0$, $q_0 = \dot{q}_0 = 1$, $q_0 = 0$ and $\dot{q}_0 = 1$, then $q_0 = 1$ and $\dot{q}_0 = 0$.

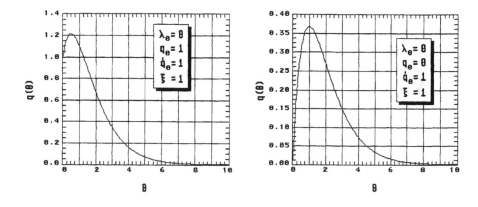

Figure 2.12. *Critical aperiodic response* $(q_0 = 0\ \dot{q}_0 = 1)$

Figure 2.13. *Critical aperiodic response* $(q_0 = 1,\ \dot{q}_0 = 1)$

$q(\theta)$ can be written $q(\theta) = \left[\dfrac{q_0}{\theta} + \left(q_0 + \dot{q}_0 \right) \right] \theta\, e^{-\theta}$. For rather large θ, $\dfrac{q_0}{\theta}$ becomes negligible and $q(\theta)$ behaves like $\left(q_0 + \dot{q}_0 \right) \theta\, e^{-\theta}$: $q(\theta)$ thus tends towards zero when θ tends towards infinity. This mode, known as *critical*, is not oscillatory. It corresponds to the fastest possible return of the system towards the equilibrium position fastest possible among all the damped exponential movements.

If we write $q_c(\theta)$ is written as the expression of $q(\theta)$ corresponding to the critical mode, this proposal can be verified while calculating:

$$\frac{q_c(\theta)}{q_{\xi>1}(\theta)}$$

Consider the expression of $q(\theta)$ [2.88] given for $\xi > 1$ in the form of a sum, the exponential terms eventually become:

$$q(\theta) \approx a\, e^{\left(\sqrt{\xi^2 - 1} - \xi\right)\theta}$$

[2.103]

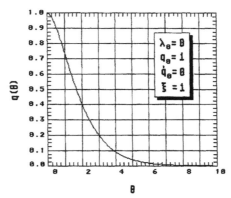

Figure 2.14. *Critical aperiodic response* $(q_0 = 1, \dot{q}_0 = 0)$

where:

$$a = \frac{\xi\left(q_0 - 2\,\lambda_0\right) + \dot{q}_0 + q_0\sqrt{\xi^2 - 1}}{2\sqrt{\xi^2 - 1}}$$

[2.104]

whereas:

$$q_c(\theta) \approx \left(q_0 + \dot{q}_0\right)\theta\, e^{-\theta}$$

[2.105]

yielding:

$$\frac{q_c(\theta)}{q(\theta)} \approx \frac{q_0 + \dot{q}_0}{a}\,\theta\, e^{-\left(1 + \sqrt{\xi^2 - 1} - \xi\right)\theta}$$

[2.106]

With the coefficient $1 - \xi + \sqrt{\xi^2 - 1}$ always being positive for $\xi > 1$, this means that the exponential term tends towards zero when θ tends towards infinity and consequently:

$$\frac{q_c(\theta)}{q(\theta)} \to 0 \qquad \qquad [2.107]$$

This return towards zero is thus performed more quickly in critical mode than in damped exponential mode.

2.6.3. *Damped oscillatory mode*

It is assumed that $0 \leq \xi < 1$.

2.6.3.1. *Free response*

The equation $p^2 + 2\,\xi\,p + 1 = 0$ has two complex roots. Let us suppose that the response is defined by an absolute movement ($\lambda_0 = 0$ for a relative movement). The response

$$q(\theta) = e^{-\xi\,\theta} \left[q_0 \cos\sqrt{1 - \xi^2}\,\theta + \frac{\dot{q} + \xi\left(q_0 - 2\,\lambda\right)}{\sqrt{1 - \xi^2}} \sin\sqrt{1 - \xi^2}\,\theta \right] \qquad [2.108]$$

can be also written:

$$q(\theta) = q_m\, e^{-\xi\,\theta} \sin\left(\sqrt{1 - \xi^2}\,\theta + \phi\right) \qquad [2.109]$$

with

$$q_m = \sqrt{q_0^2 + \frac{\left[\dot{q}_0 + \xi\left(q_0 - 2\,\lambda_0\right)\right]^2}{1 - \xi^2}} \qquad [2.110]$$

$$\tan\phi = \frac{\dot{q}_0 + \xi\left(q_0 - 2\,\lambda_0\right)}{q_0\,\sqrt{1 - \xi^2}} \qquad [2.111]$$

The response is of the damped oscillatory type with a pulsation equal to $P = \sqrt{1 - \xi^2}$, which corresponds to a period $\Theta = \dfrac{2\pi}{P}$. It is said that the movement is *damped sinusoidal* or *pseudo-sinusoidal*. The *pseudo-pulsation* P is always lower than 1. For the usual values of ξ, the pulsation is equal to 1 at first approximation $(\xi < 0.1)$.

The envelopes of the damped sinusoid have as equations:

$$q = q_m \, e^{-\xi \theta} \tag{2.112}$$

and

$$q = -q_m \, e^{-\xi \theta} \tag{2.113}$$

The free response of a mechanical system around its equilibrium position is named '*simple harmonic*'.

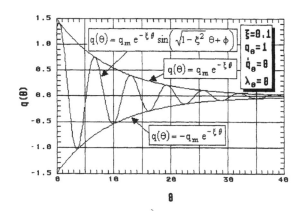

Figure 2.15. *Damped oscillatory response*

The exponent of exponential term can be written as:

$$\xi \, \theta = \xi \, \omega_0 \, t = \frac{t}{t_0}$$

t_0 is the time constant of the system and

$$t_0 = \frac{1}{\xi\,\omega_0} = \frac{\omega_0}{2\,Q} = \frac{c}{2\,m} \qquad\qquad [2.114]$$

Application

If we return to the non-reduced variables equation [2.108] can be written for the relative displacement response as:

$$u(t) = e^{-\xi\,\omega_0\,t}\left[u_0\,\cos\omega_0\,\sqrt{1-\xi^2}\;t + \frac{\dot{u}_0 + u_0\,\xi}{\sqrt{1-\xi^2}}\,\sin\omega_0\,\sqrt{1-\xi^2}\;t\right] \qquad [2.115]$$

The pseudo-pulsation is equal to

$$\omega = \omega_0\,\sqrt{1-\xi^2} \qquad\qquad [2.116]$$

with ω always being equal to or greater than ω_0.

Figure 2.16 shows the variations of the ratio $\dfrac{\omega}{\omega_0}$ with ξ.

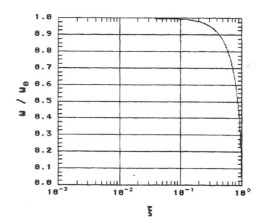

Figure 2.16. *Influence of damping on the pseudo-pulsation*

The pseudo-period

$$T = \frac{2\,\pi}{\omega_0} \qquad\qquad [2.117]$$

which separates two successive instants from the time axis crossing in the same direction, is always higher than the period of the undamped motion.

Figure 2.17 represents as an example the variations of $q(\theta)$ with θ for $\dot{q}_0$ and q_0 equal to 1 and for $\xi = 0.1$. Figures 2.18 and 2.19 show $q(\theta)$ for $\xi = 0.1$ and for $(q_0 = 1, \dot{q}_0 = 0)$ and $(q_0 = 0, \dot{q}_0 = 1)$, respectively. Figure 2.20 gives the absolute response $q(\theta)$ for $\xi = 0.1$, $q_0 = 1$, $\dot{q}_0 = 1$ and $\lambda_0 = 1$.

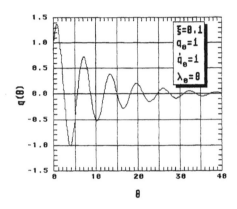

Figure 2.17. *Example of relative response for* $(q_0 = 1, \dot{q}_0 = 1)$

Figure 2.18. *Example of relative response for* $(q_0 = 1, \dot{q}_0 = 0)$

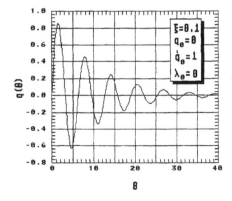
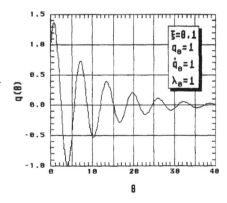

Figure 2.19. *Example of relative response for* $(q_0 = 0, \dot{q}_0 = 1)$

Figure 2.20. *Example of absolute response for* $(q_0 = 1, \dot{q}_0 = 1)$

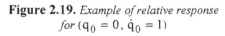

2.6.3.2. Points of contact of the response with its envelope

From $q(\theta) = q_m \, e^{-\xi\,\theta} \sin\left(\sqrt{1-\xi^2}\ \theta + \phi\right)$, the points of contact of the curve can

be determined with its envelope by seeking θ solutions of $\sin\left(\sqrt{1-\xi^2}\ \theta + \phi\right) = 1$.

These points are separated by time intervals equal to $\dfrac{\Theta}{2}$.

The points of intersection of the curve with the time axis are such that $\sin\left(P\,\theta + \phi\right) = 0$.

The maximum response is located a little before the point of contact of the curve with its envelope.

The system needs a little more than $\dfrac{\Theta}{4}$ to pass from a maximum to the next

position of zero displacement and little less than $\dfrac{\Theta}{4}$ to pass from this equilibrium

position to the position of maximum displacement.

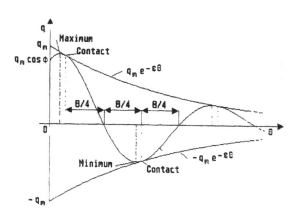

Figure 2.21. *Points of contact with the envelope*

2.6.3.3. Reduction of amplitude: logarithmic decrement

Considering two successive maximum displacements q_{1M} and q_{2M}:

$$q_{1M} = q_m \, e^{-\xi \, \theta_1} \, \sin\!\left(P \, \theta_1 + \phi\right)$$

$$q_{2M} = q_m \, e^{-\xi \, \theta_2} \, \sin\!\left(P \, \theta_2 + \phi\right)$$

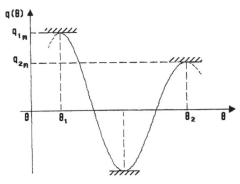

Figure 2.22. *Successive maxima of the response*

where the times θ_1 and θ_2 are such that $\dfrac{dq}{d\theta} = 0$:

$$\frac{dq}{d\theta} = q_m(-\xi) \, e^{-\xi \theta} \, \sin\!\left(P \theta + \phi\right) + q_m \, P \, e^{-\xi \theta} \, \cos\!\left(P \theta + \phi\right) \qquad [2.118]$$

$$\frac{dq}{d\theta} = 0 \text{ if}$$

$$\tan\!\left(P \, \theta + \phi\right) = \frac{P}{\xi} = \frac{\sqrt{1-\xi^2}}{\xi} \qquad [2.119]$$

i.e. if

$$\sin\!\left(P \, \theta + \phi\right) = \pm \left(\frac{\dfrac{1-\xi^2}{\xi^2}}{1 + \dfrac{1-\xi^2}{\xi^2}} \right)^{1/2} \qquad [2.120]$$

$$\sin\!\left(P \, \theta + \phi\right) = \pm \sqrt{1-\xi^2} \qquad [2.121]$$

However, $\pm\sin(P\,\theta_1 + \phi) = \pm\sin(P\,\theta_2 + \phi)$ ($\pm$ according to whether they are two maxima or two minima, but the two signs are taken to be identical), yielding

$$\frac{q_{1M}}{q_{2M}} = e^{-\xi\left(\theta_1 - \theta_2\right)}$$

[2.122]

The difference $\theta_2 - \theta_1$ is *the pseudo-period* Θ.

$$\frac{q_{1M}}{q_{2M}} = e^{\xi\,\Theta}$$

[2.123]

while $\delta = \xi\,\Theta$ is called the *logarithmic decrement*.

$$\delta = \ln\frac{q_{1M}}{q_{2M}}$$

[2.124]

This is a quantity accessible experimentally. In practice, if damping is weak, the measurement of δ is inprecise when carried out from two successive positive peaks. It is better to consider n pseudo-periods and δ is then given by [HAB 68]:

$$\boxed{\delta = \frac{1}{n}\ln\frac{q_{1M}}{q_{(n+1)M}}}$$

[2.125]

Here n is the number of positive peaks. Indeed, the ratio of the amplitude of any two consecutive peaks is [HAL 78], [LAZ 68]:

$$\frac{q_1}{q_2} = \frac{q_2}{q_3} = \frac{q_3}{q_4} = \cdots\cdots = \frac{q_n}{q_{n+1}} = e^{\delta}$$

[2.126]

yielding

$$\frac{q_1}{q_{n+1}} = \frac{q_1}{q_2}\,\frac{q_2}{q_3} = \cdots\cdots = \frac{q_n}{q_{n+1}} = e^{n\delta}$$

[2.127]

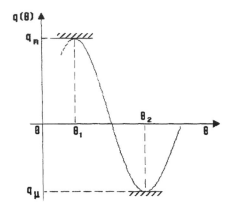

Figure 2.23. *Successive peaks of the response*

NOTES:

1. *It is useful to be able to calculate the logarithmic decrement δ starting from a maximum and a successive minimum. In this case, we can write:*

$$q_M = q_m \, e^{-\xi\theta_1} \, \sin\left(P\theta_1 + \phi\right)$$

$$q_\mu = q_m \, e^{-\xi\theta_2} \, \sin\left(P\theta_2 + \phi\right)$$

$$\sin\left(P\theta_1 + \phi\right) = -\sin\left(P\theta_2 + \phi\right)$$

$$\frac{q_M}{q_\mu} = e^{-\xi(\theta_1 - \theta_2)} = e^{\xi\frac{\Theta}{2}} = e^{\frac{\delta}{2}}$$

yielding

$$\delta = 2 \ln \left| \frac{q_M}{q_\mu} \right|$$

[2.128]

If n is even (positive or negative peaks), which corresponds to a first positive peak and the last negative peak (or the reverse), we have:

$$\delta = \frac{2}{n-1} = \ln \left| \frac{q_{1M}}{q_\mu} \right| \qquad [2.129]$$

2. The decrement δ can also be expressed according to the difference of two successive peaks:

$$\frac{q_{1M} - q_{2M}}{q_{1M}} = 1 - \frac{q_{2M}}{q_{1M}} = 1 - e^{-\delta} \qquad [2.130]$$

(indices 1 and 2 or more generally n and n + 1).

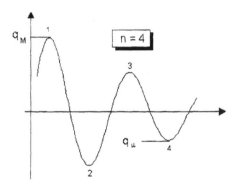

Figure 2.24. *Different peaks*

If damping is weak, q_{1M} and q_{2M} are not very different and if we set:

$$\Delta q = q_{1M} - q_{2M} \qquad [2.131]$$

Δq can be considered as infinitely small:

$$\delta = \ln \frac{q_{1M}}{q_{2M}} = \ln \left(1 + \frac{\Delta q}{q_{2M}} \right)$$

From [2.131]:

$$\delta \approx \frac{\Delta q}{q_{2M}}$$

yielding

$$\xi \approx \frac{\Delta q}{2\pi \, q_{2M}} \qquad\qquad [2.132]$$

In the case of several peaks, we have:

$$\delta \approx \frac{\Delta q}{n \, q_{2M}} \qquad\qquad [2.133]$$

and

$$\xi \approx \frac{\Delta q}{2 \, \pi \, n \, q_{2M}} \qquad\qquad [2.134]$$

Since $\dfrac{q_{1M}}{q_{2M}} = e^{\delta}$, we can connect the relative damping, ξ, and the logarithmic decrement, δ, by:

$$\xi \, \Theta = \delta \qquad\qquad [2.135]$$

Knowing that [HAB 68]:

$$\Theta = \frac{2\pi}{P} \qquad\qquad [2.136]$$

$$\boxed{\delta = \frac{2\,\pi\,\xi}{\sqrt{1 - \xi^2}}} \qquad\qquad [2.137]$$

or

$$\boxed{\xi = \frac{\delta}{\sqrt{\delta^2 + 4\,\pi^2}}} \qquad\qquad [2.138]$$

Example [LAZ 50]

Table 2.3. *Examples of decrement and damping values*

Material	δ	ξ
Concrete	0.06	0.010
Bolted steel	0.05	0.008
Welded steel	0.03	0.005

NOTE: *If ξ is very small, in practice less than 0.10, ξ^2 can, at first approximation, be neglected.*

Then:

$$\boxed{\delta \approx 2 \pi \xi}$$

[2.139]

Figure 2.25 represents the variations in the decrement δ with damping ξ and shows how, in the vicinity of the origin, one can at first approximation confuse the curve with its tangent [THO 65a].

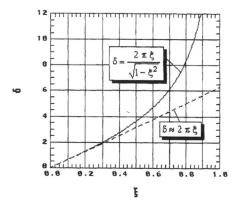

Figure 2.25. *Variations in decrement with damping*

We defined δ starting from [2.127]:

$$\frac{q_{1M}}{q_{(n+1)M}} = e^{n\,\delta}$$

yielding, by replacing δ with the expression [2.137],

$$\frac{q_{1M}}{q_{(n+1)M}} = e^{n\frac{2\pi\xi}{\sqrt{1-\xi^2}}}$$

[2.140]

i.e.

[2.141]

$$\boxed{\frac{q_{(n+1)M}}{q_{1M}} = e^{-2\pi n\xi/\sqrt{1-\xi^2}}}$$

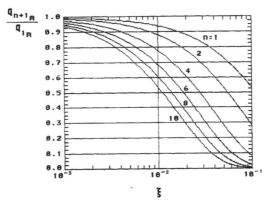

Figure 2.26. *Reduction of amplitude with damping*

The curves in Figure 2.26 give the ratio $\dfrac{q_{(n+1)M}}{q_{1M}}$ versus ξ, for various values of n. For very small ξ, we have at first approximation:

$$\frac{q_{(n+1)M}}{q_{1M}} \approx e^{-2\pi n\xi}$$

[2.142]

2.6.3.4. *Number of cycles for a given reduction in amplitude*

Amplitude reduction of 50%

On the assumption that

$$q_{(n+1)_M} = \frac{q_{1_M}}{2}$$

relation [2.125] becomes

$$\delta = \frac{1}{n} \ln 2 = \frac{2 \pi \xi}{\sqrt{1-\xi^2}}$$

[2.143]

If ξ is small:

$$2 \pi \xi \approx \frac{1}{n} \ln 2$$

$$n \xi \approx \frac{0.693}{2 \pi} \approx 0.110$$

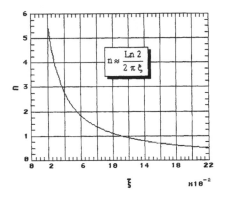

Figure 2.27. *Number of cycles for amplitude reduction of 50%*

The curve in Figure 2.27 shows the variations of n versus ξ, in the domain where the approximation $\delta \approx 2\pi\,\xi$ is correct [THO 65a].

Amplitude reduction of 90%

In the same way, we have:

$$q_{(n+1)_M} = \frac{q_{1M}}{10}$$

$$\delta = \frac{1}{n}\ln 10$$

and for small values of ξ:

$$n\,\xi \approx \frac{1}{2\pi}\ln 10$$

$$n\,\xi \approx 0.366$$

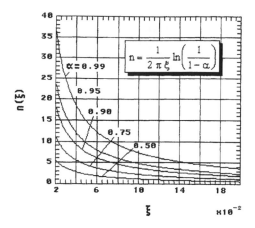

Figure 2.28. *Number of cycles for an amplitude reduction of 90%*

Reduction of α%

More generally, the number of cycles described for an amplitude reduction of α% will be, for small ξ:

$$n\,\xi \approx \frac{1}{2\pi}\,\ln\left[\frac{1}{1-\dfrac{\alpha}{100}}\right] \qquad\qquad [2.144]$$

2.6.3.5. Influence of damping on period

Except if it is very large, damping in general has little influence over the period; we have $\Theta_0 = \dfrac{2\pi}{Po}$ for $\xi = 0$ ($P_0 = 1$) and, for small ξ,

$$\frac{\Delta\Theta}{\Theta_0} = \frac{\Theta - \Theta_0}{\Theta_0} = \frac{\Theta}{\Theta_0} - 1 = \frac{\dfrac{2\pi}{P}}{\dfrac{2\pi}{P_0}} - 1 = \frac{P_0}{P} - 1 = \frac{1}{\sqrt{1-\xi^2}} - 1 \approx \frac{\delta^2}{8\,\pi^2} \qquad [2.145]$$

For most current calculations, it is possible to confuse Θ with Θ_0. For the first order, the pulsation and the period are not modified by damping. For the second order, the pulsation is modified by a corrective term that is always negative and the period is increased:

$$\Theta = \frac{\Theta_0}{\sqrt{1 - \xi^2}} \approx \Theta_0 \left(1 + \frac{\xi^2}{2} \right)$$

[2.146]

and

$$P = P_0 \sqrt{1 - \xi^2} \approx P_0 \left(1 - \frac{\xi^2}{2} \right)$$

[2.147]

$\left(P_0 = 1 \right)$ or

$$\omega = \omega_0 \sqrt{1 - \xi^2} \approx \omega_0 \left(1 - \frac{\xi^2}{2} \right)$$

[2.148]

NOTE: *The logarithmic decrement also represents the energy variation during a cycle of decrease. For sufficiently small δ we have [LAZ 50]:*

$$\delta \approx \frac{1}{2} \left(\frac{\text{Energy}(n-1) - \text{Energy}(n)}{\text{Energy}(n-1)} \right)$$

[2.149]

$$\left| \text{Energy}(n-1) = \text{Energy to the } (n-1)^{\text{th}} \text{ cycle} \right|$$

In practical cases where ξ lies between 0 and 1, the energy initially provided to the system dissipates itself little by little in the external medium in various forms (friction between solid bodies, with air or another fluid, internal slips in the metal during elastic strain, radiation, energy dissipation in electromagnetic form).

Consequently, the amplitude of the oscillations decreases constantly with time. If we wanted to keep a constant amplitude, it would be necessary to restore the system with the energy which it loses every time. The system is then no longer free: the oscillations are maintained or forced. We will study this case in Chapter 5.

2.6.3.6. *Particular case of zero damping*

In this case, $q(\theta)$ becomes:

$$q(\theta) = q_0 \cos \theta + \dot{q}_0 \sin \theta$$

[2.150]

which can be also written as:

$$q(\theta) = q_m \sin (\theta + \varphi)$$ [2.151]

where

$$q_m = \sqrt{q_0^2 + \dot{q}_0^2}$$ [2.152]

$$\text{tg } \varphi = \frac{\dot{q}_0}{q_0}$$ [2.153]

If it is supposed that the mechanical system has moved away from its equilibrium position and then is released in the absence of any external forces at time $t = 0$, the response is then of the non-damped oscillatory type for $\xi = 0$.

In this (theoretical) case, the movement of natural pulsation ω_0 should last indefinitely since the characteristic equation does not contain a first order term. This is the consequence of the absence of a damping element. The potential energy of the spring decreases by increasing the kinetic energy of the mass and vice versa: the system is known as *conservative*.

In summary, when the relative damping ξ varies continuously, the mode passes without discontinuity to one of the following:

$\xi = 0$ Undamped oscillatory mode.

$0 \leq \xi \leq 1$ Damped oscillatory mode. The system moves away from its equilibrium position, oscillating around the equilibrium point before stabilizing.

$\xi = 1$ Critical aperiodic mode, corresponding to the fastest possible return of the system without crossing the equilibrium position.

$\xi > 1$ Damped aperiodic mode. The system returns to its equilibrium position without any oscillation, all the more quickly because ξ is closer to the unit.

In the following chapters, we will focus more specifically on the case $0 \leq \xi \leq 1$, which corresponds to the values observed in the majority of real structures.

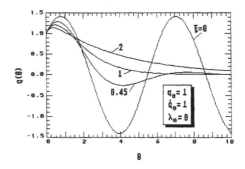

Figure 2.29. *Various modes for* $q_0 = 1$, $\dot{q}_0 = 1$

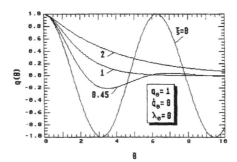

Figure 2.30. *Various modes for* $q_0 = 1$. $\dot{q}_0 = 0$

2.6.3.7. *Quality factor*

The *quality factor* or *Q factor* of the oscillator is the number Q defined by:

$$\frac{1}{Q} = \frac{c}{2\,\omega_0} = \frac{c}{\sqrt{k\,m}} = \frac{\omega_0\,c}{k} = 2\,\xi$$ [2.154]

The properties of this factor will be considered in more detail in the following chapters.

Chapter 3

Impulse and step responses

3.1. Response of a mass–spring system to a unit step function (step or indicial response)

3.1.1. *Response defined by relative displacement*

3.1.1.1. *Expression for response*

Let us consider a damped mass–spring system. Before the initial time $t = 0$ the mass is assumed to be at rest. At time $t = 0$, a constant excitation of unit amplitude continuously acts for all $t > 0$ [BRO 53], [KAR 40]. We have seen that, for zero initial conditions the Laplace transform of the response of a one-degree-of-freedom system is given by [2.29]:

$$Q(p) = \frac{\Lambda(p)}{p^2 + 2\,\xi\,p + 1} \tag{3.1}$$

Here $\Lambda(p) = \dfrac{1}{p}$ (unit step transform); yielding the response:

$$q(\theta) = L^{-1}\left[\frac{1}{p\left(p^2 + 2\,\xi\,p + 1\right)}\right] = L^{-1}\left[\frac{1}{p}\right] - L^{-1}\left[\frac{p}{p^2 + 2\,\xi\,p + 1}\right] - L^{-1}\left[\frac{2\,\xi}{p^2 + 2\,\xi\,p + 1}\right] \tag{3.2}$$

$$q(\theta) = 1 - \frac{e^{-\xi\,\theta}}{\sqrt{1-\xi^2}}\left[\sqrt{1-\xi^2}\,\cos\sqrt{1-\xi^2}\,\theta - \xi\sin\sqrt{1-\xi^2}\,\theta\right]$$

$$-2\,\xi\,\frac{e^{-\xi\,\theta}}{\sqrt{1-\xi^2}}\sin\sqrt{1-\xi^2}\,\theta \qquad\qquad [3.3]$$

$(\xi \neq 1)$

$$q(\theta) = 1 - e^{-\xi\,\theta}\left[\cos\sqrt{1-\xi^2}\,\theta + \frac{\xi}{\sqrt{1-\xi^2}}\sin\sqrt{1-\xi^2}\,\theta\right] \qquad [3.4]$$

i.e.

$$u(t) = A(t) = \ell_m\left[1 - e^{-\xi\,\omega_0\,t}\cos\omega_0\sqrt{1-\xi^2}\,t - \frac{\xi}{\sqrt{1-\xi^2}}e^{-\xi\,\omega_0\,t}\sin\omega_0\sqrt{1-\xi^2}\,t\right]$$

$$[3.5]$$

with $\ell_m = 1$ [HAB 68], [KAR 40].

NOTE: *This calculation is identical to that carried out to obtain the primary response spectrum to a rectangular shock [LAL 75].*

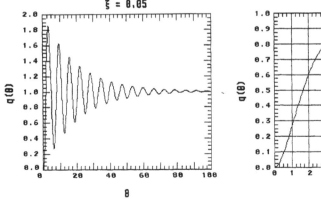

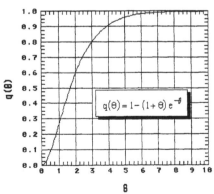

Figure 3.1. *Example of relative displacement response to a unit step excitation*

Figure 3.2. *Response to a unit step excitation for $\xi = 1$*

Specific cases

1. If $\xi = 1$

$$Q(p) = \frac{1}{p\,(p+1)^2} \qquad\qquad [3.6]$$

$$Q(p) = \frac{1}{p} - \frac{1}{p+1} - \frac{1}{(p+1)^2} \qquad\qquad [3.7]$$

$$q(\theta) = 1 - (1+\theta)\,e^{-\theta} \qquad\qquad [3.8]$$

and, for $\ell_m = 1$,

$$u(t) = 1 - e^{-\omega_0 t} - \omega_0\,t\,e^{-\omega_0 t} \qquad\qquad [3.9]$$

2. Zero damping

In the reduced form, the equation of movement can be written using the notation of the previous paragraphs:

$$\frac{d^2 q(\theta)}{d\theta^2} + q(\theta) = \lambda(\theta), \qquad\qquad [3.10]$$

or

$$\ddot{u}(t) + \omega_0^2\,u(t) = \omega_0^2\,\ell(t) \qquad\qquad [3.11]$$

with\the initial conditions being constant, namely, for $\theta = 0$

$$q(0) = \left(\frac{dq}{d\theta}\right)_{\theta=0} = 0$$

or, according to the case, $t = 0$ and

$$u(0) = \left(\frac{du}{dt}\right)_{t=0} = 0.$$

After integration, this becomes as before:

$$q(\theta) = 1 - \cos\theta \qquad\qquad [3.12]$$

and

$$u(t) = \ell_m \left(1 - \cos \omega_0 t\right)$$ [3.13]

the expression in which, by definition of the excitation, $\ell_m = 1$:

$$u(t) = 1 - \cos \omega_0 t$$ [3.14]

Example

If the excitation is a force, the equation of the movement is, for $t \geq 0$,

$$m\frac{d^2 z}{dt^2} + k\, z = 1, \text{ with, for initial conditions at } t = 0, \quad z(0) = \left(\frac{dz}{dt}\right)_{t=0} = 0.$$

yielding, after integration,

$$z(t) = \frac{1}{k}\left(1 - \cos\sqrt{\frac{k}{m}}\, t\right)$$ [3.15]

NOTE: *The dimensions of [3.15] do not seem correct. It should be remembered that the excitation used is a force of amplitude unit equal to* $\dfrac{1}{k}$ *which is thus homogeneous with a displacement.*

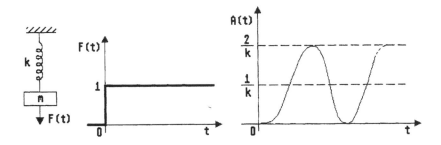

Figure 3.3. *Step response for* $\xi = 0$

The function $z(t)$, response with the step unit function, is often termed *indicial admittance* or *step response* and is written $A(t)$.

It can be seen in this example that if we set, according to our notation, $z_s = \dfrac{F_m}{k} = \dfrac{1}{k}$, the ratio of the maximum elongation z_m to the static deflection z_s which the mass would take if the force were statically applied reached a value of 2. The spring, in dynamics, performs twice more often than in statics, although it is unlikely to undergo stresses that are twice as large. Often, however, the materials resist the transient stresses better than static stresses (Chapter 1). This remark relates to the first instants of time during which $F(t)$ is transitory and is raised from 0 to 1. For this example where $F(t)$ remains equal to one for all t positive values and where the system is undamped, the effect of shock would be followed by a fatigue effect.

3.1.1.2. *Extremum for response*

The expression for the response

$$q(\theta) = 1 - e^{-\xi\theta}\left[\cos\sqrt{1-\xi^2}\,\theta + \frac{\xi}{\sqrt{1-\xi^2}}\sin\sqrt{1-\xi^2}\,\theta\right] \qquad [3.16]$$

has a zero derivative $\dfrac{dq}{d\theta}$ for $\theta = \theta_m$ such that

$$-\xi\,e^{-\xi\theta_m}\left[\cos\sqrt{1-\xi^2}\,\theta_m + \frac{\xi}{\sqrt{1-\xi^2}}\sin\sqrt{1-\xi^2}\,\theta_m\right]$$

$$+ e^{-\xi\theta_m}\left[-\sqrt{1-\xi^2}\sin\sqrt{1-\xi^2}\,\theta_m + \xi\cos\sqrt{1-\xi^2}\,\theta_m\right] = 0$$

$$\sin\sqrt{1-\xi^2}\,\theta_m = 0$$

$$\theta_m = \frac{k\,\pi}{\sqrt{1-\xi^2}} \qquad [3.17]$$

The first maximum (which would correspond to the point of the positive primary shock response spectrum at the natural frequency f_0 of the resonator) occurs for

$$\theta_m = \frac{\pi}{\sqrt{1-\xi^2}} \quad \text{at time } t_m = \frac{1}{2\,f_0\sqrt{1-\xi^2}} \quad \text{[HAL 78]}.$$

From this the value $q(\theta)$ is deduced:

$$q(\theta_m) = q_m = 1 - e^{-\dfrac{\xi\pi}{\sqrt{1-\xi^2}}} \left[\cos\pi + \frac{\xi}{\sqrt{1-\xi^2}} \sin\pi \right] \qquad [3.18]$$

$$q_m = 1 + e^{-\dfrac{\pi\xi}{\sqrt{1-\xi^2}}} \qquad [3.19]$$

(always a positive quantity). The first maximum amplitude q_m tends towards 1 when ξ tends towards 1.

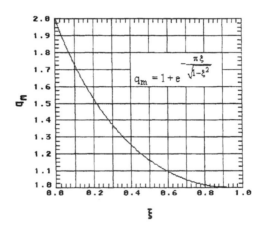

Figure 3.4. *First maximum amplitude versus ξ*

NOTES:

1. q_m *is independent of the natural frequency of the resonator.*

2. *For $\xi = 0$, $q_m = 2$.*

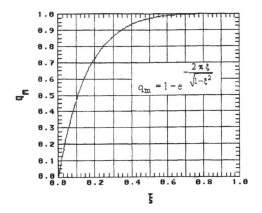

Figure 3.5. *Amplitude of the first minimum versus ξ*

For k = 2,

$$\theta_m = \frac{2\pi}{\sqrt{1-\xi^2}},$$ [3.20]

and

$$t_m = \frac{1}{f_0\sqrt{1-\xi^2}},$$

$$q(\theta_m) = q_m = 1 - e^{\dfrac{2\pi\xi}{\sqrt{1-\xi^2}}}$$ [3.21]

q_m is negative for all $\xi \in [0, 1]$.

$q_m = 0$ for $\xi = 0$

$q_m = 1$ for $\xi = 1$

3.1.1.3. *First excursion of response to unit value*

θ_1 is searched such that:

$$q(\theta) \equiv 1 = 1 - e^{-\xi\,\theta_1} \left[\cos\sqrt{1-\xi^2}\;\theta_1 + \frac{\xi}{\sqrt{1-\xi^2}}\sin\sqrt{1-\xi^2}\;\theta_1 \right]$$ [3.22]

As $e^{-\xi\,\theta_1} \neq 0$ is assumed

$$\cos\sqrt{1-\xi^2}\ \theta_1 = -\frac{\xi}{\sqrt{1-\xi^2}}\ \sin\sqrt{1-\xi^2}\ \theta_1$$

i.e.

$$\tan\sqrt{1-\xi^2}\ \theta_1 = -\frac{\sqrt{1-\xi^2}}{\xi} \qquad\qquad [3.23]$$

This yields, since $\tan\sqrt{1-\xi^2}\ \theta_1 \le 0$ and $\sqrt{1-\xi^2}\ \theta_1 \ge 0$ must be present simultaneously:

$$\sqrt{1-\xi^2}\ \theta_1 = \pi - \mathrm{Arc\,tan}\,\frac{\sqrt{1-\xi^2}}{\xi} \qquad\qquad [3.24]$$

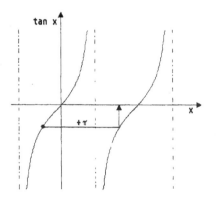
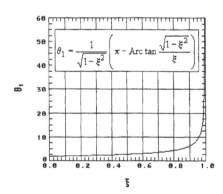

Figure 3.6. *Resolution of [3.23]*   Figure 3.7. *Time of first excursion by the unit step response*

$$\boxed{\theta_1 = \frac{1}{\sqrt{1-\xi^2}}\left[\pi - \mathrm{Arc\,tan}\,\frac{\sqrt{1-\xi^2}}{\xi}\right]} \qquad\qquad [3.25]$$

If $\xi = 0$,

$$q(\theta) = 1 - \cos\theta \qquad [3.26]$$

If $q(\theta) = 1$

$$\theta = \left(k + \frac{1}{2}\right)\pi \qquad [3.27]$$

If $\xi = 1$

$$1 = 1 - e^{-\theta} - \theta e^{-\theta} \qquad [3.28]$$

The only positive root exists for infinite θ.

3.1.2. Response defined by absolute displacement, velocity or acceleration

3.1.2.1. *Expression for response*

In this case, for any ξ and zero initial conditions,

$$Q(p) = \frac{\Lambda(p)\left(1 + 2\xi p\right)}{p^2 + 2\xi p + 1} \qquad [3.29]$$

with $\Lambda(p) = \dfrac{1}{p}$.

$$q(\theta) = L^{-1}\left[\frac{1 + 2\xi p}{p\left(p^2 + 2\xi p + 1\right)}\right] = L^{-1}\left[\frac{1}{p}\right] - L^{-1}\left[\frac{p}{p^2 + 2\xi p + 1}\right] \qquad [3.30]$$

$$\boxed{q(\theta) = 1 - e^{-\xi\theta}\left[\cos\sqrt{1 - \xi^2}\ \theta - \frac{\xi}{\sqrt{1 - \xi^2}}\sin\sqrt{1 - \xi^2}\ \theta\right] = A(\theta)} \qquad [3.31]$$

$(\xi \neq 1)$.

If $\xi = 0$

$$q(\theta) = 1 - \cos\theta \qquad [3.32]$$

If $\xi = 1$

$$Q(p) = \frac{1}{p} \frac{1+2p}{(p+1)^2}$$ [3.33]

$$Q(p) = \frac{1}{p} - \frac{1}{p+1} + \frac{1}{(p+1)^2}$$ [3.34]

$$q(\theta) = 1 - e^{-\theta} + \theta e^{-\theta} = \Lambda(\theta) = 1 + (\theta - 1) e^{-\theta}$$ [3.35]

$$u(t) = 1 + (\omega_0 t - 1) e^{-\omega_0 t}$$ [3.36]

$(\ell_m = 1)$.

3.1.2.2. Extremum for response

The extremum of the response $q(\theta) = A(\theta)$ occurs for $\theta = \theta_m$ such that $\dfrac{dA}{d\theta} = 0$, which leads to

$$-\xi e^{-\xi \theta_m} \left[\cos \sqrt{1 - \xi^2}\, \theta_m - \frac{\xi}{\sqrt{1 - \xi^2}} \sin \sqrt{1 - \xi^2}\, \theta_m \right]$$

$$+ e^{-\xi \theta_m} \left[-\sqrt{1 - \xi^2} \sin \sqrt{1 - \xi^2}\, \theta_m - \xi \cos \sqrt{1 - \xi^2}\, \theta_m \right] = 0$$

i.e. to

$$\tan \sqrt{1 - \xi^2}\, \theta_m = \frac{2\xi \sqrt{1 - \xi^2}}{2\xi^2 - 1}$$ [3.37]

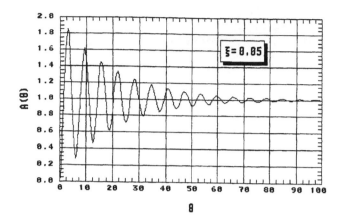

Figure 3.8. *Example of absolute response*

For $2\,\xi^2 - 1 \ge 0$ (and since θ_m is positive):

$$\theta_m = \frac{1}{\sqrt{1-\xi^2}}\ \mathrm{Arc\,tan}\ \frac{2\,\xi\,\sqrt{1-\xi^2}}{2\,\xi^2 - 1} \qquad\qquad [3.38]$$

and if $2\,\xi^2 - 1 < 0$

$$\theta_m = \frac{1}{\sqrt{1-\xi^2}}\left[\pi + \mathrm{Arc\,tan}\ \frac{2\,\xi\,\sqrt{1-\xi^2}}{2\,\xi^2 - 1}\right] \qquad\qquad [3.39]$$

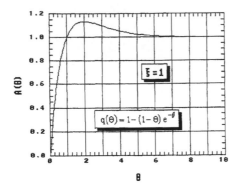

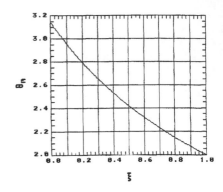

Figure 3.9. *Amplitude of the absolute response for* $\xi = 1$

Figure 3.10. *Time of extremum versus* ξ

For $\theta = \theta_m$,

$$A(\theta_m) = 1 - e^{-\xi\,\theta_m}\left[\cos\sqrt{1-\xi^2}\,\theta_m - \frac{\xi}{\sqrt{1-\xi^2}}\sin\sqrt{1-\xi^2}\,\theta_m\right]$$

i.e.

$$A(\theta_m) = 1 + e^{-\xi\,\theta_m}$$ [3.40]

If $\xi = 1$,

$$q(\theta) = 1 - e^{-\theta} + \theta\,e^{-\theta}$$ [3.41]

Then

$$\frac{dq}{d\theta} = (2 - \theta)\,e^{-\theta} = 0$$

if $\theta = 2$ or if $\theta \to \infty$.

It yields $q(\theta_m) = 1 + e^{-2}$

or

$$q(\theta_m) = 1$$

If $\xi = \dfrac{1}{\sqrt{2}}$, $\theta_m = \dfrac{\pi}{2}\sqrt{2}$ and

$$q(\theta_m) = 1 + e^{-\pi/2}$$

$$q(\theta) \approx 1.20788\ldots$$

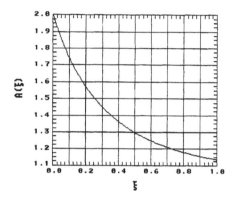

Figure 3.11. *Amplitude of absolute response versus* ξ

3.1.2.3. *First passage of the response by the unit value*

The first excursion of the unit value happens at time θ_1 such that

$$A(\theta_1) = 1 - e^{-\xi\,\theta_1}\left[\cos\sqrt{1-\xi^2}\;\theta_1 - \frac{\xi}{\sqrt{1-\xi^2}}\sin\sqrt{1-\xi^2}\;\theta_1\right] \qquad [3.42]$$

$$\tan\sqrt{1-\xi^2}\;\theta_1 = \frac{\sqrt{1-\xi^2}}{\xi} \qquad [3.43]$$

$$\boxed{\theta_1 = \frac{1}{\sqrt{1-\xi^2}}\,\text{Arc}\tan\frac{\sqrt{1-\xi^2}}{\xi}} \qquad [3.44]$$

If $\xi = 0$,

$$\theta_1 = \frac{\pi}{2}$$

If $\xi = 1$,

$$q(\theta) = 1 = 1 - e^{-\theta} + \theta \, e^{-\theta} \qquad\qquad [3.45]$$

yielding $\theta_1 = 1$ or $\theta_1 =$ infinity.

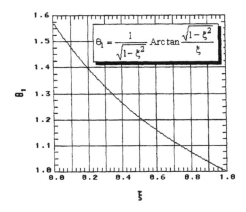

Figure 3.12. *Time of first transit of the unit value versus ξ*

3.2. Response of a mass–spring system to a unit impulse excitation

3.2.1. *Response defined by relative displacement*

3.2.1.1. *Expression for response*

Let us consider a Dirac delta function $\delta_g(\theta)$ obeying

$$\left| \begin{array}{ll} \delta_g(\theta) = 0 & \text{for } \theta \neq 0 \\ \delta(0) & \text{infinite} \\ \int_{-\infty}^{+\infty} \delta(\theta) \, d\theta = 1 \end{array} \right. \qquad\qquad [3.46]$$

such that, with $\theta = 0$, $q = 0$ and $m \dfrac{dq}{dt} = 1$. The quantity $m \dfrac{dq}{dt}$ is the impulse transmitted to the mass m by a force acting for a small interval of time $\Delta\theta$ [KAR 40]. The contribution of the restoring force of the spring to the impulse is negligible during the very short time interval $\Delta\theta$.

Depending on whether the impulse is defined by a force or an acceleration, then

$$\frac{d^2z}{dt^2} + 2\,\xi\,\omega_0\,\frac{dz}{dt} + \omega_0^2\,z = \omega_0^2\,\frac{\delta_F}{k} \qquad [3.47]$$

or

$$\frac{d^2z}{dt^2} + 2\,\xi\,\omega_0\,\frac{dz}{dt} + \omega_0^2\,z = -\omega_0^2\,\frac{\delta_{AC}}{\omega_0^2} \qquad [3.48]$$

If $\delta_g(t)$ the generalized delta function is equal, according to the case, to $\dfrac{\delta_F}{k}$ or

to $-\dfrac{\delta_{AC}}{\omega_0^2}$, then the generalized equation is obtained as follows:

$$\boxed{\frac{d^2u}{dt^2} + 2\,\xi\,\omega_0\,\frac{du}{dt} + \omega_0^2\,u = \omega_0^2\,\delta_g(t)} \qquad [3.49]$$

Then

$$\Im = \int_0^{\Delta t} \delta_g(t)\,dt = \begin{cases} \displaystyle\int_0^{\Delta t} \frac{\delta_F}{k}\,dt = \frac{1}{k}\,I \\[4mm] \displaystyle\int_0^{\Delta t}\left(-\frac{\delta_{AC}}{\omega_0^2}\right) dt = -\frac{1}{\omega_0^2}\,I \end{cases} \qquad [3.50]$$

($I = 1$). To make the differential equation dimensionless each member is divided by the quantity $\Im\,\omega_0$ homogeneous with length and set $q = \dfrac{u}{\Im\,\omega_0}$ and $\theta = \omega_0\,t$. This becomes:

$$\frac{d^2q}{d\theta^2} + 2\,\xi\,\frac{dq}{d\theta} + q(\theta) = \delta_g(\theta) \qquad [3.51]$$

The Laplace transform of this equation is written with the notation already used,

$$Q(p)\left(p^2 + 2\,\xi\,p + 1\right) = 1 \qquad\qquad [3.52]$$

and since the transform of a Dirac delta function is equal to the unit [LAL 75], this gives

$$\boxed{q(\theta) = \frac{e^{-\xi\,\theta}}{\sqrt{1-\xi^2}}\,\sin\sqrt{1-\xi^2}\,\theta = h(\theta)} \qquad\qquad [3.53]$$

($\xi \neq 1$) and

$$u(t) = h(t) = \omega_0\,\Im\,\frac{e^{-\xi\,\omega_0\,t}}{\sqrt{1-\xi^2}}\,\sin\omega_0\,\sqrt{1-\xi^2}\,t \qquad\qquad [3.54]$$

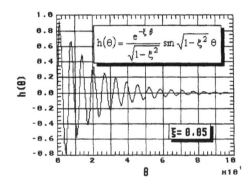

Figure 3.13. *Impulse response*

Specific cases

1. For $\xi = 0$,

$$h(\theta) = \sin\theta \qquad\qquad [3.55]$$

and

$$u(t) = \omega_0\,\Im\,\sin\omega_0\,t \qquad\qquad [3.56]$$

Example

If the impulse is defined by a force, $\Im = \dfrac{1}{k}$ then

$$u(t) = z(t) = \frac{\omega_0}{k} \sin \omega_0 \, t = \frac{1}{\sqrt{k \, m}} \sin \omega_0 \, t \qquad\qquad [3.57]$$

This relation is quite homogeneous, since the 'number' 1 corresponds to the impulse $\dfrac{1}{\sqrt{k \, m}} = \dfrac{1}{\sqrt{k \, m}}$, which has the dimension of a displacement.

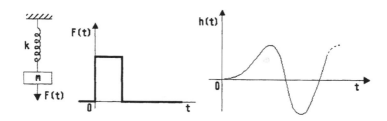

Figure 3.14. *Impulse response*

The unit impulse response is denoted by $h(t)$ [BRO 53] [KAR 40]. It is termed *impulse response, impulsive response* or *weight function* [GUI 63].

2. If $\xi = 1$,

$$Q(p) = \frac{1}{(p+1)^2} \qquad\qquad [3.58]$$

$$q(\theta) = h(\theta) = \theta \, e^{-\theta} \qquad\qquad [3.59]$$

$$u(t) = h(t) = \omega_0^2 \, t \, e^{-\omega_0 \, t} \qquad\qquad [3.60]$$

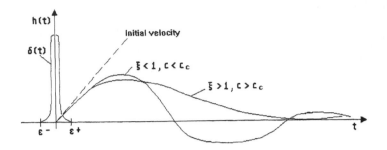

Figure 3.15. *Examples of impulse responses versus* ξ

3.2.1.2. *Extremum for response*

$q(\theta)$ presents a peak q_m when $\dfrac{dq}{d\theta} = 0$. i.e. for $\theta = \theta_m$ such that

$$-\xi\, e^{-\xi\,\theta_m} \sin \sqrt{1-\xi^2}\ \theta_m + e^{-\xi\,\theta_m} \sqrt{1-\xi^2}\ \cos \sqrt{1-\xi^2}\ \theta_m = 0$$

$$\tan \sqrt{1-\xi^2}\ \theta_m = \frac{\sqrt{1-\xi^2}}{\xi}$$

$$\theta_m = \frac{1}{\sqrt{1-\xi^2}}\ \text{Arc tan} \frac{\sqrt{1-\xi^2}}{\xi} \qquad [3.61]$$

This yields

$$q_m = \frac{e^{-\xi\,\theta_m}}{\sqrt{1-\xi^2}}\ \sin \sqrt{1-\xi^2}\ \theta_m$$

i.e.

$$q_m = e^{-\frac{\xi}{\sqrt{1-\xi^2}}\text{Arc tan}\frac{\sqrt{1-\xi^2}}{\xi}} \qquad [3.62]$$

For $\xi = 1$,

$$q(\theta) = \theta\, e^{-\theta} \qquad [3.63]$$

$$\frac{dq}{d\theta} = 0 \text{ if } \theta = 1, \text{ yielding}$$

$$q_m = \frac{1}{e}$$ [3.64]

For $\xi = 0$,

$$h(\theta) = \sin \theta$$ [3.65]

$$\frac{dh}{d\theta} = \cos \theta = 0 \text{ if } \theta = \left(k + \frac{1}{2}\right)\pi. \text{ If } k = 0, h = 1.$$

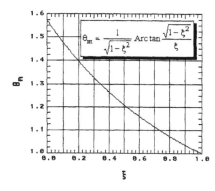

Figure 3.16. *Time of the first maximum versus ξ*

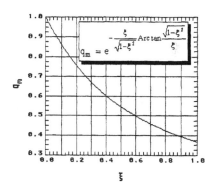

Figure 3.17. *Amplitude of the first maximum versus ξ*

NOTE: *The comparison of the Laplace transforms of the response with the unit step function*

$$Q(p) = \frac{1}{p} \frac{1}{p^2 + 2\xi p + 1}$$ [3.66]

and the unit impulse

$$Q(p) = \frac{1}{p^2 + 2\xi p + 1}$$ [3.67]

shows that these two transforms differ by a factor $\dfrac{1}{p}$ *and that, consequently [BRO 53], [KAR 40],*

$$h(t) = \frac{dA(t)}{dt} \tag{3.68}$$

3.2.2. Response defined by absolute parameter

3.2.2.1. *Expression for response*

$$Q(p) = \frac{1 + 2\,\xi\,p}{p^2 + 2\,\xi\,p + 1} \tag{3.69}$$

$$q(\theta) = h(\theta) = \frac{e^{-\xi\,\theta}}{\sqrt{1-\xi^2}}\sin\sqrt{1-\xi^2}\,\theta + 2\,\xi\,e^{-\xi\,\theta}\left[\cos\sqrt{1-\xi^2}\,\theta - \frac{\xi}{\sqrt{1-\xi^2}}\sin\sqrt{1-\xi^2}\,\theta\right] \tag{3.70}$$

$(\xi \neq 1)$

$$h(\theta) = e^{-\xi\,\theta}\left[2\,\xi\,\cos\sqrt{1-\xi^2}\,\theta + \frac{1-2\,\xi^2}{\sqrt{1-\xi^2}}\sin\sqrt{1-\xi^2}\,\theta\right] \tag{3.71}$$

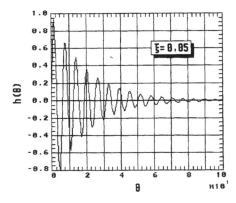

Figure 3.18. *Absolute response*

NOTE: *For* $\theta = 0$,

$$h(0) = 2\,\xi = \frac{1}{Q}$$
[3.72]

If $\xi = 0$, the preceding case is found

$$h(\theta) = \sin\,\theta$$
[3.73]

In non-reduced coordinates the impulse response is written [BRO 62]:

$$h(t) = \omega_0\,\Im\,e^{-\xi\,\omega_0\,t}\left[2\,\xi\cos\omega_0\sqrt{1-\xi^2}\,t + \frac{1-2\,\xi^2}{\sqrt{1-\xi^2}}\sin\omega_0\sqrt{1-\xi^2}\,t\right]$$
[3.74]

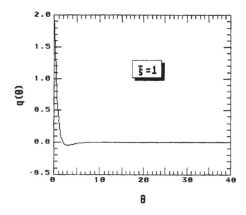

Figure 3.19. *Absolute response for $\xi = 1$*

If $\xi = 1$,

$$Q(p) = \frac{1+2\,p}{(p+1)^2} = \frac{2}{p+1} - \frac{1}{(p+1)^2}$$
[3.75]

$$q(\theta) = 2\,e^{-\theta} - \theta\,e^{-\theta} = (2-\theta)\,e^{-\theta}$$
[3.76]

$$h(t) = \Im\,\omega_0\left(2 - \omega_0\,t\right)e^{-\omega_0\,t}$$
[3.77]

3.2.2.2. *Peaks of response*

The response $h(\theta)$ presents a peak when $\dfrac{dh}{d\theta} = 0$, i.e. for $\theta = \theta_m$ such that

$$- \xi e^{-\xi \theta_m} \left[2\xi \cos \sqrt{1 - \xi^2}\ \theta_m + \frac{1 - 2\xi^2}{\sqrt{1 - \xi^2}} \sin \sqrt{1 - \xi^2}\ \theta_m \right]$$

$$+ e^{-\xi \theta_m} \left[- 2\xi \sqrt{1 - \xi^2} \sin \sqrt{1 - \xi^2}\ \theta_m + \left(1 - 2\xi^2\right) \cos \sqrt{1 - \xi^2}\ \theta_m \right] = 0$$

i.e., after simplification, if

$$\tan\left(\theta_m \sqrt{1 - \xi^2} \right) = \frac{\sqrt{1 - \xi^2}\left(1 - 4\xi^2\right)}{\xi\left(3 - 4\xi^2\right)} \tag{3.78}$$

If $3 - 4\xi^2 < 0$ (i.e. $\xi > \dfrac{\sqrt{3}}{2}$).

$$\theta_m = \frac{1}{\sqrt{1 - \xi^2}} \operatorname{Arc\,tan} \left[\frac{\sqrt{1 - \xi^2}}{\xi} \frac{1 - 4\xi^2}{3 - 4\xi^2} \right] \tag{3.79}$$

and if $3 - 4\xi^2 \geq 0$, i.e. $\xi \leq \dfrac{\sqrt{3}}{2}$,

$$\theta_m = \frac{1}{\sqrt{1 - \xi^2}} \left[\pi + \operatorname{Arc\,tan} \left(\frac{\sqrt{1 - \xi^2}}{\xi} \frac{1 - 4\xi^2}{3 - 4\xi^2} \right) \right] \tag{3.80}$$

this yields

$$h(\theta_m) = h_m = e^{-\xi \theta_m} \left[2\xi \cos \sqrt{1 - \xi^2}\ \theta_m + \frac{1 - 2\xi^2}{\sqrt{1 - \xi^2}} \sin \sqrt{1 - \xi^2}\ \theta_m \right] \tag{3.81}$$

i.e.

$$h_m = -e^{-\xi \theta_m} \tag{3.82}$$

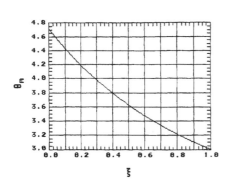

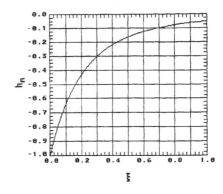

Figure 3.20. *Time of the first maximum of the absolute response versus ξ*   **Figure 3.21.** *Amplitude of the first maximum of the absolute response versus ξ*

If $\xi = 1$.

$$\frac{dq}{d\theta} = -2\,e^{-\theta} - e^{-\theta} + \theta\,e^{-\theta} = 0 \qquad\qquad [3.83]$$

Since $e^{-\theta} \neq 0$, we obtain $\theta_m = 3$ and $h_m = e^{-3} \approx -0.049\,788\ldots$

If $\xi = 0$,

$$h(\theta) = \sin\theta \qquad\qquad [3.84]$$

$\dfrac{dh}{d\theta} = \cos\theta = 0$ if $\theta_m = \pi\left(k + \dfrac{1}{2}\right)$. If $k = 0$, $\theta_m = \dfrac{\pi}{2}$ and $h_m = 1$.

NOTE:

1. *The equation [2.37], which can be written*

$$u(t) = \frac{\omega_0}{\sqrt{1-\xi^2}} \int_0^t \ell(\alpha)\,e^{-\xi\,\omega_0\,(t-\alpha)} \sin\omega_0\sqrt{1-\xi^2}\,(t-\alpha)\,d\alpha.$$

is none other than a convolution integral applied to the functions $\ell(t)$ and

$$h(t) = \frac{\omega_0}{\sqrt{1-\xi^2}}\,e^{-\xi\,\omega_0\,t}\sin\omega_0\sqrt{1-\xi^2}\,t \qquad\qquad [3.85]$$

(h(t) = *impulse response or weight function*).

2. *The Fourier transform of a convolution product of two functions ℓ and H is equal to the product of their Fourier transforms [LAL 75]. If* $u = \ell * h$

$$U(\Omega) = TF(U) = TF(\ell * h) = L(\Omega).H(\Omega)$$ [3.86]

The function $H(\Omega)$, *Fourier transform of the impulse response, is the transfer function of the system [LAL 75].*

3. *In addition, the Laplace transformation applied to a linear one-degree-of-freedom differential equation leads to a similar relation:*

$$U(p) = A(p) L(p)$$ [3.87]

$A(p)$ *is termed* operational admittance *and* $Z(p) = \dfrac{1}{A(p)}$ generalized impedance *of the system.*

4. *In the same way, the relation [2.60] can be considered as the convolution product of the two functions* $\ell(t)$ *and*

$$h(t) = \frac{\omega_0}{\sqrt{1-\xi^2}} \ e^{-\xi\,\omega_0\,t}\left[\left(1 - 2\,\xi^2\right)\sin\omega_0\sqrt{1-\xi^2}\ t + 2\,\xi\,\sqrt{1-\xi^2}\ \cos\omega_0\,\sqrt{1-\xi^2}\ t\right]$$

[3.88]

3.3. Use of step and impulse responses

The preceding results can be used to calculate the response of the linear one-degree-of-freedom system (k, m) to an arbitrary excitation $\ell(t)$. This response can be considered in two ways [BRO 53]:

– either as the sum of the responses of the system to a succession of impulses of very short duration (the envelope of these impulses corresponding to the excitation) (Figure 3.22);

– or as the sum of the responses of the system to a series of step functions (Figure 3.23).

The application of the superposition principle supposes that the system is linear, i.e. described by linear differential equations [KAR 40] [MUS 68].

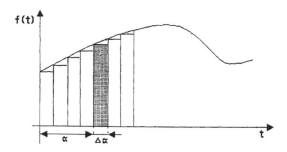

Figure 3.22. *Arbitrary pulse as a series of impulses*

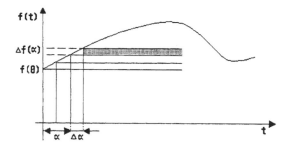

Figure 3.23. *Arbitrary pulse as a series of step functions*

Let us initially regard the excitation $\ell(t)$ as a succession of very short duration $\Delta\alpha$ impulses and let $\ell(\alpha)$ be the impulse amplitude at time α. By hypothesis, $\ell(\alpha) = 0$ for $\alpha < 0$.

Set $h(t - \alpha)$ as the response of the system at time t, resulting from the impulse at the time α pertaining to the time interval (0, t) (paragraph 3.2.1). The response $z(t)$ of the system to all the impulses occurring between $\alpha = 0$ and $\alpha = t$ is:

$$u(t) = \sum_{\alpha=0}^{\alpha=t} \ell(\alpha)\, h(t - \alpha)\, \Delta\alpha \qquad\qquad [3.89]$$

If the excitation is a continuous function, the intervals $\Delta\alpha$ can tend towards zero; it then becomes (Duhamel's formula):

$$u(t) = \int_0^t \ell(\alpha)\, h(t - \alpha)\, d\alpha \qquad\qquad [3.90]$$

The calculation of this integral requires knowledge of the excitation function $\ell(t)$ and of the response $h(t - \alpha)$ to the unit impulse at time α.

The integral [3.90] is none other than a convolution integral [LAL 75]; this can then be written as:

$$\ell(t) * h(t) = \int_0^t \ell(\alpha)\, h(t - \alpha)\, d\alpha \qquad\qquad [3.91]$$

According to properties of the convolution [LAL 75]:

$$\ell(t) * h(t) = \int_0^t \ell(t - \alpha)\, h(\alpha)\, d\alpha \qquad\qquad [3.92]$$

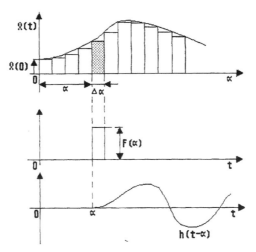

Figure 3.24. *Summation of impulse responses*

NOTE: *It is supposed in this calculation that at time t the response to an impulse applied at time α is observed, so that this response is only one function of the time interval $t - \alpha$, but not of t or of α separately. This is the case if the coefficients of the differential equation of the system are constant. This assumption is in general not justified if these coefficients are functions of time [KAR 40].*

Consider the excitation as a sum of step functions separated by equal time intervals $\Delta\alpha$ (Figure 3.25).

The amplitude of each step function is $\Delta\ell(\alpha)$, i.e. $\dfrac{\Delta\ell(\alpha)}{\Delta\alpha}\,\Delta\alpha$. Set $A(t-\alpha)$ as the step response at time t, resulting from the application of a unit step function at time α (with $0 < \alpha < t$).

Set $\ell(0)$ as the value of the excitation at time $\alpha = 0$ and $A(t)$ as the response of the system at time t corresponding to the application of the unit step at the instant $\alpha = 0$.

The response of the system to a single unit step function is equal to

$$\frac{\Delta\ell(\alpha)}{\Delta\alpha}\,\Delta\alpha\;A(t-\alpha)$$

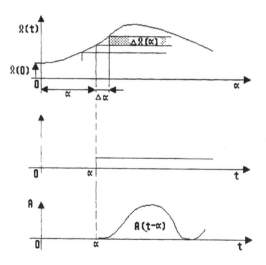

Figure 3.25. *Summation of step responses*

The response of the linear system to all the step functions applied between the times $\alpha = 0$ and $\alpha = t$ and separated by $\Delta\alpha$ is thus:

$$u(t) = \ell(0) \cdot A(t) + \sum_{\alpha=0}^{\alpha=t} \frac{\Delta \ell(\alpha)}{\Delta \alpha} \, \Delta \alpha \, A(t - \alpha) \qquad [3.93]$$

If the excitation function is continuous, the response tends, when $\Delta \alpha$ tends towards zero, towards the limit

$$u(t) = \ell(0) \cdot A(t) + \int_0^t \dot\ell(\alpha) \, A(t - \alpha) \, d\alpha \qquad [3.94]$$

where

$$\dot\ell(\alpha) = \frac{d\ell(\alpha)}{d\alpha}$$

This is the *superposition integral* or *Rocard integral*. In the majority of practical cases, and according to our assumptions, $\ell(0) = 0$ and

$$u(t) = \int_0^t \dot\ell(0) \, A(t - \alpha) \, d\alpha \qquad [3.95]$$

NOTES:

1. *The expression [3.95] is sometimes called Duhamel's integral and sometimes [3.90] Rocard's integral [RID 69].*

2. *The integral [3.90] can be obtained by the integration by parts of [3.94] while setting* $U = A(t - \alpha)$ *and* $dV = \dot\ell(\alpha) \, d\alpha$ *using integration by parts, by noting that* $A(0)$ *is often zero in most current practical problems (knowing moreover that* $h(t) = \dfrac{dA(t)}{dt}$ *).*

$$u(t) = \ell(0) \cdot A(t) + \int_0^t \dot\ell(\alpha) \, A(t - \alpha) \, d\alpha \qquad [3.96]$$

If u is a continuous and derivable function in (0, t), integration by parts gives

$$\int_0^t \dot\ell(\alpha) \, A(t - \alpha) \, d\alpha = \left[\ell(\alpha) \, A(t - \alpha) \right]_0^t - \int_0^t \ell(\alpha) \, \frac{dA(t - \alpha)}{d\alpha} \, d\alpha \qquad [3.97]$$

yielding, if $\ell_0 = \ell(0)$,

$$u(t) = \ell_0 \, A(t) + \ell(t) \, A(0) - \ell_0 \, A(t) + \int_0^t \ell(\alpha) \, \dot A(t - \alpha) \, d\alpha \qquad [3.98]$$

and, since $A(0) = 0$, by Duhamel's formula:

$$u(t) = \int_0^t \ell(\alpha)\, \dot{A}(t - \alpha)\, d\alpha \qquad\qquad [3.99]$$

The functions $h(t)$ and $A(t)$ were calculated directly in the preceding paragraphs. Their expression could be obtained by starting from the general equation of movement in its reduced form. The next step will be to find, for example, $h(t)$. The unit impulse can be defined, in generalized form, by the integral:

$$\lim_{\theta \to 0} \int_0^\theta \lambda(\alpha)\, d\alpha = 1 \qquad\qquad [3.100]$$

α being a variable of integration ($\alpha \le \theta$). This relation defines an excitation where the duration is infinitely small and whose integral in time domain is equal to the unit. Since it corresponds to an excitation of duration tending towards zero, it can be regarded as an initial condition to the solution of the equation of motion

$$\ddot{q}(\theta) + q(\theta) = \lambda(\theta) \qquad\qquad [3.101]$$

(while supposing $\xi = 0$) i.e.

$$q(\theta) = C_1 \cos\theta + C_2 \sin\theta \qquad\qquad [3.102]$$

The initial value of the response $q(\theta)$ is equal to C_1 and, for a system initially at rest ($C_1 = 0$), the initial velocity is C_2. The amplitude of the response being zero for $\theta = 0$, the initial velocity change is obtained by setting $q = 0$ in the equation of movement [3.101], while integrating $\dot{q} = \dfrac{dq}{d\theta}$ over time and taking the limit when θ tends towards zero [SUT 68]:

$$\dot{q}(\theta \to 0) = \lim_{\theta \to 0}\left(\int_0^\theta \frac{d\dot{q}}{d\alpha}\, d\alpha \right) = \lim_{\theta \to 0}\left(\int_0^\theta \lambda(\alpha)\, d\alpha \right) \qquad\qquad [3.103]$$

yielding

$$C_2 = 1$$

This then gives the expression of the response to the generalized unit impulse of an undamped simple system:

$$q(\theta) = \sin\theta \qquad\qquad [3.104]$$

– for zero damping, the indicial admittance and the impulse response to the generalized excitation are written, respectively:

$$A(t) = 1 - \cos \omega_0 t \qquad\qquad\qquad [3.105]$$

and

$$h(t) = \omega_0 \sin \omega_0 t \qquad\qquad\qquad [3.106]$$

This yields

$$u(t) = \int_0^t \ell(\alpha)\, h(t - \alpha)\, d\alpha \qquad\qquad\qquad [3.107]$$

$$u(t) = \omega_0 \int_0^t \ell(\alpha)\, \sin \omega_0 (t - \alpha)\, d\alpha \qquad\qquad\qquad [3.108]$$

for arbitrary ξ damping,

$$A(t) = 1 - e^{-\xi \omega_0 t} \cos \omega_0 \sqrt{1 - \xi^2}\, t - \frac{\xi}{\sqrt{1 - \xi^2}} e^{-\xi \omega_0 t} \sin \omega_0 \sqrt{1 - \xi^2}\, t$$

$$[3.109]$$

and

$$h(t) = \frac{\omega_0}{\sqrt{1 - \xi^2}} e^{-\xi \omega_0 t} \sin \omega_0 \sqrt{1 - \xi^2}\, t \qquad\qquad\qquad [3.110]$$

yielding

$$u(t) = \frac{\omega_0}{\sqrt{1 - \xi^2}} \int_0^t \ell(\alpha)\, e^{-\xi \omega_0 (t-\alpha)} \sin \omega_0 \sqrt{1 - \xi^2}\, (t - \alpha)\, d\alpha \qquad [3.111]$$

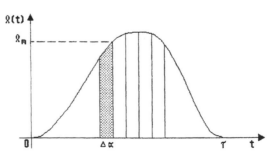

Figure 3.26. *Decline in impulses*

The response of the simple system of natural pulsation ω_0 can therefore be calculated after a decline of the excitation $\ell(t)$ in a series of impulses of duration $\Delta\alpha$. For a signal of given form, the displacement $u(t)$ is a function of t, ω_0 and ξ.

3.4. Transfer function of a linear one-degree-of-freedom system

3.4.1. *Definition*

It was shown in [3.90] that the behaviour of a linear system can be characterized by its weight function (response of the system to an unit impulse function)

$$q(\theta) = \int_0^\theta \lambda(\alpha) \, h(\theta - \alpha) \, d\alpha$$

where, if the response is relative,

$$\left[\begin{array}{ll} h(\theta) = \dfrac{e^{-\xi\theta}}{\sqrt{1-\xi^2}} \sin\sqrt{1-\xi^2}\,\theta & (\xi \neq 1) \\[4mm] h(\theta) = \theta\, e^{-\theta} & (\xi = 1) \end{array} \right. \tag{3.112}$$

and, if it is absolute:

$$\left[\begin{array}{ll} h(\theta) = e^{-\xi\theta}\left[2\,\xi \cos\sqrt{1-\xi^2}\,\theta + \dfrac{1-2\xi^2}{\sqrt{1-\xi^2}} \sin\sqrt{1-\xi^2}\,\theta \right] & \text{for } \xi \neq 1 \\[4mm] h(\theta) = (2-\theta)\,e^{-\theta} & \text{for } \xi = 1 \end{array} \right. \tag{3.113}$$

The function $h(\ )$ can be expressed versus time. We have, for example, for the relative response:

$$h(t) = \frac{\omega_0}{\sqrt{1-\xi^2}} e^{-\xi\omega_0 t} \sin\sqrt{1-\xi^2}\,\omega_0 t \tag{3.114}$$

The Fourier transform of $h(t)$ is the transfer function $H(\Omega)$ of the system [BEN 63]:

$$\boxed{H(\Omega) = \int_0^\infty h(t) \, e^{-i\Omega t} dt} \tag{3.115}$$

Let us set $h = \dfrac{\Omega}{\omega_0}$. The variable h is defined as the *interval*. In reduced coordinates:

$$H(h) = \int_0^\infty h(\theta)\, e^{-i h \theta}\, d\theta$$ [3.116]

NOTE: *Rigorously,* $H(h)$ *is the response function in the frequency domain, whereas the transfer function is the Laplace transform of* $h(\theta)$ *[KIM 24]. Commonly,* $H(h)$ *is also known as the transfer function.*

The function $H(h)$ [1] is complex and can be put in the form [BEN 63]

$$H(h) = |H(h)|\, e^{-i \phi(h)}$$ [3.117]

Sometimes the module $|H(h)|$ is called the *gain factor* [KIM 24] or *gain*, or *amplification factor* when $h(\theta)$ is the relative response function; or *transmissibility* when $h(\theta)$ is the absolute response function and $\phi(h)$ is the associated phase (*phase factor*).

Taking into account the characteristics of real physical systems, $H(h)$ satisfies the following properties:

1. $\qquad H(-h) = H^*(h)$ [3.118]

where H^* is the complex conjugate of H

2. $\qquad |H(-h)| = |H(h)|$ [3.119]

3. $\qquad \phi(-h) = -\phi(h)$ [3.120]

4. If two mechanical systems having transfer functions $H_1(h)$ and $H_2(h)$ are put in series and if there is no coupling between the two systems thus associated, the transfer function of the unit is equal to [BEN 63]:

$$H(h) = H_1(h)\, H_2(h)$$ [3.121]

i.e.

---

[1] The dimensionless term 'h' is used throughout this and succeeding chapters. This is equivalent to the frequency ratios f / f_0 or ω / ω_0

$$\begin{cases} |H(h)| = |H_1(h)| \cdot |H_2(h)| \\ \phi(h) = \phi_1(h) + \phi_2(h) \end{cases}$$
[3.122]

This can be found in references [LAL 75] [LAL 82] [LAL 95a] and examples of use of this transfer function for the calculation of the response of a structure in a given point when it is subjected to a sinusoidal, random or shock excitation are given in the following chapters.

In a more general way, the transfer function can be defined as the ratio of the response of a structure (with several degrees of freedom) to the excitation, according to the frequency. The stated properties of $H(h)$ remain valid with this definition. The function $H(h)$ depends only on the structural characteristics.

3.4.2. Calculation of $H(h)$ for relative response

By definition,

$$H(h) = \int_0^\infty \frac{e^{-\xi\theta}}{\sqrt{1-\xi^2}} \sin\left(\sqrt{1-\xi^2}\,\theta\right) e^{-ih\theta}\, d\theta$$
[3.123]

Knowing that

$$\int e^{ax} \sin bx\, dx = \frac{e^{ax}}{a^2 + b^2}\left(a \sin bx - b \cos bx\right)$$
[3.124]

it becomes

$$H(h) = \frac{1}{\sqrt{1-\xi^2}} \left\{ \frac{e^{-(\xi+ih)}}{1-\xi^2 + (\xi+ih)^2} \right.$$

$$\left. \left[-(\xi+ih)\sin\sqrt{1-\xi^2}\,\theta - \sqrt{1-\xi^2}\,\cos\sqrt{1-\xi^2}\,\theta \right] \right\}_0^\infty$$
[3.125]

$$H(h) = \frac{1}{\left(1-h^2\right) + 2\,i\,h\,\xi}$$
[3.126]

$$\boxed{|H(h)| = \frac{1}{\sqrt{\left(1-h^2\right)^2 + 4\,\xi^2\,h^2}}}$$ [3.127]

$$\operatorname{tg}\phi = \frac{2\,\xi\,h}{1-h^2}$$ [3.128]

If $0 \le h < 1$:

$$\boxed{\phi = \operatorname{Arc\,tan}\frac{2\,\xi\,h}{1-h^2}}$$ [3.129]

If $h = 1$

$$\phi = \frac{\pi}{2}$$ [3.130]

If $h > 1$

$$\phi = \pi + \operatorname{Arc\,tan}\frac{2\,\xi\,h}{1-h^2}$$ [3.131]

3.4.3. *Calculation of* $H(h)$ *for absolute response*

In this case,

$$H(h) = \int_0^\infty e^{-\xi\theta}\left[2\,\xi\cos\sqrt{1-\xi^2}\,\theta + \frac{1-2\,\xi^2}{\sqrt{1-\xi^2}}\sin\sqrt{1-\xi^2}\,\theta\right]e^{-\imath\,h\,\theta}\,d\theta$$

[3.132]

$$H(h) = \int_0^\infty 2\,\xi\,e^{-(\xi+\imath\,h)\,\theta}\cos\sqrt{1-\xi^2}\,\theta\,d\theta + \frac{1-2\,\xi^2}{\sqrt{1-\xi^2}}\int_0^\infty e^{-(\xi+\imath\,h)\,\theta}\sin\sqrt{1-\xi^2}\,\theta\,d\theta$$

[3.133]

$$H(h) = \left\{ 2\xi \frac{e^{-(\xi+ih)\theta}}{(\xi+ih)^2 + 1 - \xi^2} \left[-(\xi+ih) \cos\sqrt{1-\xi^2}\ \theta + \sqrt{1-\xi^2} \sin\sqrt{1-\xi^2}\ \theta\ d\theta \right] \right.$$

$$\left. + \frac{1-2\xi^2}{\sqrt{1-\xi^2}} \frac{e^{-(\xi+ih)\theta}}{(\xi+ih)^2 + 1 - \xi^2} \left[-(\xi+ih) \sin\sqrt{1-\xi^2}\ \theta - \sqrt{1-\xi^2} \cos\sqrt{1-\xi^2}\ \theta \right] \right\}_0^{\infty}$$

[3.134]

$$H(h) = \frac{2\xi(\xi+ih)}{1-h^2+2i\xi h} + \frac{(1-2\xi^2)\sqrt{1-\xi^2}}{\sqrt{1-\xi^2}\left(1-h^2+2i\xi h\right)}$$

[3.135]

$$H(h) = \frac{1+2i\xi h}{1-h^2+2i\xi h} \left(= \frac{1-h^2+4h^2\xi^2-2i\xi h^3}{\left(1-h^2\right)^2+4\xi^2 h^2} \right)$$

[3.136]

$$\left| H(h) \right| = \frac{\sqrt{1+4h^2\xi^2}}{\sqrt{\left(1-h^2\right)^2+4\xi^2 h^2}}$$

[3.137]

$$\tan\phi = \frac{2\xi h^3}{1-h^2+4h^2\xi^2}$$

[3.138]

$$\phi = \text{Arc tan} \frac{2\xi h^3}{1-h^2+4h^2\xi^2}$$

[3.139]

if $1-h^2+4h^2\xi^2 > 0$, i.e. if $h^2 < \dfrac{1}{1-4\xi^2}$. For $h^2 = \dfrac{1}{1-4\xi^2}$,

$$\phi = \frac{\pi}{2}$$

[3.140]

and for $h^2 > \dfrac{1}{1-4\xi^2}$

$$\phi = \pi + \text{Arc tan} \ \frac{2 \ \xi \ h^3}{1 - h^2 + 4 \ \xi^2 \ h^2} \qquad [3.141]$$

$$\text{If } \xi = \frac{1}{2}, \ \tan \phi = h^3$$

$$\phi = \text{Arc tan } h^3 \qquad [3.142]$$

The complex transfer function can also be studied through its real and imaginary parts (Nyquist diagram):

$$H(f) = \frac{1 + i \ 2 \ \xi \ h}{1 - h^2 + i \ 2 \ \xi \ h} = \text{Re}[H(f)] + i \ \text{Im}[H(f)] \qquad [3.143]$$

$$\text{Re}[H(f)] = \frac{1 - h^2 + (2 \ \xi \ h)^2}{(1 - h^2)^2 + (2 \ \xi \ h)^2} \qquad [3.144]$$

$$\text{Im}[H(f)] = \frac{-2 \ \xi \ h^3}{(1 - h^2)^2 + (2 \ \xi \ h)^2} \qquad [3.145]$$

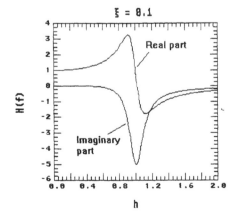

Figure 3.27. *Real and imaginary parts of* H(h)

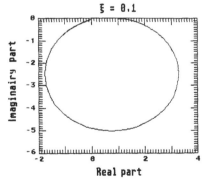

Figure 3.28. *Nyquist diagram*

3.4.4. *Other definitions of transfer function*

3.4.4.1. *Notation*

According to choice of the parameters for excitation and response, the transfer function can be defined differently. In order to avoid any confusion, the two letters are placed in as subscripts after the letter H; the first letter specifies the nature of the input, the second that of the response. The letter H will be used without indeces only in the case of reduced coordinates. We will use the same rule for the impedance $\dfrac{1}{H} = Z$.

3.4.4.2. *Relative response*

$$|H| = \frac{1}{\sqrt{\left(1 - h^2\right)^2 + \left(2\,\xi\,h\right)^2}} \qquad [3.146]$$

The function $|H|$ is equal to the function $\left|H_{\ell,\,u}\right| = \dfrac{u}{\ell}$. To distinguish it from the transfer function giving the absolute responses we will denote it by H_R.

Calculation of $\left|H_{\ddot{x},\,z}\right|$

$$|\ell(t)| = \frac{\ddot{x}(t)}{\omega_0^2} \qquad [3.147]$$

$$\frac{u}{\ell} = \frac{z}{\ddot{x}}\,\omega_0^2 \qquad [3.148]$$

$$\left|H_{\ddot{x},\,z}\right| = \left|H_{\ell,\,u}\right|\frac{1}{\omega_0^2} \qquad [3.149]$$

Calculation of $\left|H_{F,\,z}\right|$

$$\frac{u}{\ell} = \frac{z}{F\,/\,k} \qquad [3.150]$$

$$\left|H_{F,z}\right| = \frac{\left|H_{\ell.u}\right|}{k}$$ [3.151]

Response defined by the relative velocity

$$\left|H_{\ell,\dot{u}}\right| = \frac{\dot{u}}{\ell} = \frac{\Omega\, u}{\ell}$$ [3.152]

It is supposed here that the excitation, and consequently the response, are sinusoidal and of frequency Ω, or that the excitation is resoluble into a Fourier series, with each component being a sinusoid. This yields

$$\left|H_{\ell,\dot{u}}\right| = \Omega\,\left|H_{\ell,u}\right| = \Omega\,\left|H\right|$$ [3.153]

Thus, if $\left|\ell(t)\right| = \dfrac{\ddot{x}(t)}{\omega_0^2}$,

$$\frac{\dot{u}}{\ell} = \frac{\dot{z}}{\ddot{x}/\omega_0^2} = \Omega\,\left|H\right|$$ [3.154]

$$H_{\ddot{x},z} = \frac{\Omega}{\omega_0^2}\,\left|H\right|$$ [3.155]

3.4.4.3. *Absolute response*

In the same way, starting from [3.137],

$$\left|H\right| = \sqrt{\frac{1 + 4\,\xi^2\, h^2}{\left(1 - h^2\right)^2 + \left(2\,\xi\, h\right)^2}}$$

noting H_A, it is possible to calculate all the expressions of the usual transfer functions of this nature.

3.4.4.4. *Summary tables*

Table 3.1 states the values of H_A and H_R for each parameter input and each parameter response.

Example

Suppose that the excitation and the response are, respectively, the velocities $\dot{x}$ and $\dot{z}$. Table 3.1 indicates that the transfer function can be obtained from the relation

$$H_R = \frac{\omega_0^2\,\dot{z}}{\Omega^2\,\dot{x}}$$

[3.156]

Yielding

$$\frac{\dot{z}}{\dot{x}} = \frac{\Omega^2}{\omega_0^2}\frac{1}{\sqrt{\left(1-h^2\right)^2 + 4\,\xi^2\,h^2}}$$

[3.157]

Table 3.2 gives this relation more directly. To continue to use reduced parameters, and in particular of reduced transfer functions (which is not the case for the transfer functions in Table 3.2), these functions can be defined as follows.

Table 3.1. *Transfer function corresponding to excitation and response*

Response ⇒ Excitation ⇓	z	$\dot{z}$	$\ddot{z}$	y	$\dot{y}$	$\ddot{y}$	Reaction force F_T on base
Force F on the mass m	$\dfrac{k\,z}{F}$	$\dfrac{k\,\dot{z}}{\Omega\,F}$	$\dfrac{k\,\ddot{z}}{\Omega^2\,F}$	/	/	/	$\dfrac{F_T}{F}$
$\ddot{x}$	$\dfrac{\omega_0^2\,z}{\ddot{x}}$	$\dfrac{\omega_0^2\,\dot{z}}{\Omega\,\ddot{x}}$	$\dfrac{\omega_0^2\,\ddot{z}}{\Omega^2\,\ddot{x}}$	$\dfrac{\Omega^2\,y}{\ddot{x}}$	$\dfrac{\Omega\,\dot{y}}{\ddot{x}}$	$\dfrac{\ddot{y}}{\ddot{x}}$	/
$\dot{x}$	$\dfrac{\omega_0^2\,z}{\Omega\,\dot{x}}$	$\dfrac{\omega_0^2\,\dot{z}}{\Omega^2\,\dot{x}}$	$\dfrac{\omega_0^2\,\ddot{z}}{\Omega^3\,\dot{x}}$	$\dfrac{\Omega\,y}{\dot{x}}$	$\dfrac{\dot{y}}{\dot{x}}$	$\dfrac{\ddot{y}}{\Omega\,\dot{x}}$	/
x	$\dfrac{\omega_0^2\,z}{\Omega^2\,x}$	$\dfrac{\omega_0^2\,\dot{z}}{\Omega^3\,x}$	$\dfrac{\omega_0^2\,\ddot{z}}{\Omega^4\,x}$	$\dfrac{y}{x}$	$\dfrac{\dot{y}}{\Omega\,x}$	$\dfrac{\ddot{y}}{\Omega^2\,x}$	/
Reduced transfer function	H_R			H_A			

These results are also be presented as in Table 3.2.

For a given excitation, we obtain the acceleration and *velocity* and *acceleration* transfer function while multiplying respectively by h and h^2 the *displacement* transfer function (relative or absolute response).

This is used to draw the transfer function in a four coordinate representation from which can be read (starting from only one curve plotted against the reduced frequency h) the transfer function for the displacement, velocity and acceleration (paragraph 5.7).

NOTE: *The transfer functions are sometimes expressed in decibels*

$$H(dB) = 20 \log_{10} H(h)$$ [3.158]

where $H(h)$ *is the amplitude of the transfer function as defined in the preceding tables. A variation of H(h) by a factor of 10 corresponds to an amplification of 20 dB.*

Table 3.2. *Transfer function corresponding to excitation and response*

Response ⇒ Excitation ⇓		Displacement (m)		Velocity (m/s)		Acceleration (m/s²)		Reaction force on the base
		Absolute $y(t)$	Relative $z(t)$	Absolute $\dot{y}(t)$	Relative $z(t)$	Absolute $\ddot{y}(t)$	Relative $\ddot{z}(t)$	/
Base movement	Displacement $x(t)$ (m)	H_A	$\dfrac{\Omega^2}{\omega_0^2} H_R$	ΩH_A	$\dfrac{\Omega^3}{\omega_0^2} H_R$	$\Omega^2 H_A$	$\dfrac{\Omega^4}{\omega_0^2} H_A$	/
	Velocity $\dot{x}(t)$ (m/s)	$\dfrac{H_A}{\Omega}$	$\dfrac{\Omega}{\omega_0^2} H_R$	H_A	$\dfrac{\Omega^2}{\omega_0^2} H_R$	ΩH_A	$\dfrac{\Omega^3}{\omega_0^2} H_R$	/
	Acceleration $\ddot{x}(t)$ (m/s²)	$\dfrac{H_A}{\Omega^2}$	$\dfrac{H_R}{\omega_0^2}$	$\dfrac{H_A}{\Omega}$	$\dfrac{\Omega}{\omega_0^2} H_R$	H_A	$\dfrac{\Omega^2}{\omega_0^2} H_R$	/
Force on the mass m		$\dfrac{H_R}{k}$		$\dfrac{\Omega}{k} H_R$		$\dfrac{\Omega^2}{k} H_R$		H_A

Table 3.3. *Transfer function corresponding to excitation and response*

Response → / Excitation ⇓	Displacement Absolute $y(t)$	Displacement Relative $z(t)$	Velocity Absolute $\dot{y}(t)$	Velocity Relative $\dot{z}(t)$	Acceleration Absolute $\ddot{y}(t)$	Acceleration Relative $\ddot{z}(t)$	Reaction force $F(t)$ on the base
Base movement — Displacement $x(t)$	$\dfrac{y}{x}=H_A$	$\dfrac{z}{x}=h^2\,H_R$	$\dfrac{\dot{y}}{\omega_0\,x}=h\,H_A$	$\dfrac{\dot{z}}{x}=h^3\,H_R$	$\dfrac{\ddot{y}}{\omega_0^2\,x}=h^2\,H_A$	$\dfrac{\ddot{z}}{\omega_0^2\,x}=h^4\,H_R$	/
Velocity $\dot{x}(t)$	$\dfrac{\omega_0\,y}{\dot{x}}=\dfrac{H_A}{h}$	$\dfrac{\omega_0\,z}{\dot{x}}=h\,H_R$	$\dfrac{\dot{y}}{\dot{x}}=H_A$	$\dfrac{\dot{z}}{\dot{x}}=h^2\,H_R$	$\dfrac{\ddot{y}}{\omega_0\,\dot{x}}=h\,H_A$	$\dfrac{\ddot{z}}{\omega_0\,\dot{x}}=h^3\,H_R$	/
Acceleration $\ddot{x}(t)$	$\dfrac{\omega_0^2\,y}{\ddot{x}}=\dfrac{H_A}{h^2}$	$\dfrac{\omega_0^2\,z}{\ddot{x}}=H_R$	$\dfrac{\omega_0\,\dot{y}}{\ddot{x}}=\dfrac{H_A}{h}$	$\dfrac{\omega_0\,\dot{z}}{\ddot{x}}=h\,H_R$	$\dfrac{\ddot{y}}{\ddot{x}}=H_A$	$\dfrac{\ddot{z}}{\ddot{x}}=h^2\,H_R$	/
Force $F(t)$ on the mass m	$\dfrac{k\,z}{F}=H_R$		$\dfrac{\sqrt{k\,m}\,\dot{z}}{F}=h\,H_R$		$\dfrac{m\,\ddot{z}}{F}=h^2\,H_R$		$\dfrac{F_T}{F}=H_A$

3.5. Measurement of transfer function

The transfer function $H(f)$ of a mechanical system can be defined:

– in steady state sinusoidal mode, by calculation of the amplitude ratio of the response to the amplitude of the excitation for several values of the frequency f of the excitation [TAY 77];

– in a slowly swept sine, the sweep rate being selected as sufficiently slow so that the transient aspect can be neglected when crossing the resonances;

– in a quickly swept sine (method developed by C.W. Skingle [SKI 66]);

– under random vibrations (the ratio of the power spectral density functions of the response and excitation, or the ratio of the cross-spectral density $G_{\ddot{x}\ddot{y}}$ and power spectral density of the excitation $G_{\ddot{x}}$) (see Volume 3);

– under shock (ratio of the Fourier transforms of the response and excitation) (see Volume 2). In this last case, a hammer equipped with a sensor measuring the input force and a sensor measuring acceleration response or, as with the preceding methods, an electrodynamic shaker can be used.

Most of the authors agree that the fast swept sine is one of the best methods of measurement of the transfer function of a system. Shock excitation can give good results provided that the amplitude of the Fourier transform of the shock used has a level far enough from zero in all the useful frequency bands. The random vibrations require longer tests [SMA 85] [TAY 75].

Chapter 4

Sinusoidal vibration

4.1. Definitions

4.1.1. *Sinusoidal vibration*

A vibration is known as *sinusoidal* when it can be described analytically by an expression of the form

$$\ell(t) = \ell_m \sin(\Omega t + \varphi) \tag{4.1}$$

where

$\ell(t)$ is the parameter used to define the excitation. $\ell(t)$ is in general an acceleration, but can be a velocity, a displacement (linear or angular) or a force;

t is the instantaneous value of time (seconds);

φ is the phase (related to the value of ℓ for $t = 0$). φ is expressed in radians. In practice, it is supposed that $\varphi = 0$ if it is possible.

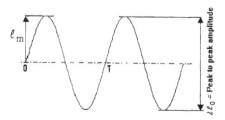

Figure 4.1. *Sinusoidal vibration*

The sinusoid is characterized by:

– its frequency f (expressed in Hertz) or of its period $T = \dfrac{1}{f}$ (in seconds) or of its pulsation $\Omega = 2\,\pi\,f$ (rad/sec);

– its duration;

– its amplitude defined by the maximum value ℓ_m of $\ell(t)$ or by the peak-to-peak amplitude $2\,\ell_m$.

Excitation can be defined by an acceleration, a velocity or a displacement, these three parameters resulting one from the other by integration or by derivation:

$$\begin{cases} \dot{\ell}(t) = \dfrac{d\ell}{dt} = \ell_m\,\Omega\,\cos\Omega\,t = \dot{\ell}_m\,\cos\Omega\,t = \dot{\ell}_m\,\sin\!\left(\Omega\,t + \dfrac{\pi}{2}\right) \\[2mm] \ddot{\ell}(t) = \dfrac{d^2\ell}{dt^2} = -\ell_m\,\Omega^2\,\sin\Omega\,t = -\ddot{\ell}_m\,\sin\Omega\,t = \ddot{\ell}_m\,\sin(\Omega\,t + \pi) \end{cases}$$

[4.2]

From these expressions, it can be observed that acceleration, the velocity and the displacement are all sinusoidal, of period T, and that the velocity and the displacement have a difference of phase angle of $\dfrac{\pi}{2}$, as well as the velocity and acceleration.

4.1.2. *Mean value*

The mean value of the quantity $\ell(t)$, which is defined over one period T by

$$\bar{\ell} = \frac{1}{T}\int_0^T \ell(t)\,dt$$

[4.3]

is zero (over one period, there is symmetry of all the points with respect to the time axis, to within a translation of $\dfrac{T}{2}$). The surface under the positive part (between the curve and the time axis) is equal to the surface under the negative part. The average value of the signal on a half-period is more significant:

$$\overline{\ell} = \frac{2}{T} \int_0^{T/2} \ell(t) \, dt \tag{4.4}$$

$$\overline{\ell} = \frac{2}{T} \ell_m \int_0^{T/2} \sin \Omega t \, dt$$

(yielding, since $\Omega T = 2\pi$)

$$\overline{\ell} = \frac{2 \ell_m}{\pi} \approx 0.637 \, \ell_m \tag{4.5}$$

4.1.3. *Mean square value – rms value*

The *mean square value* is defined as

$$\overline{\ell^2} = \frac{1}{T} \int_0^T \ell^2(t) \, dt \tag{4.6}$$

$$\overline{\ell^2} = \frac{1}{T} \int_0^T \ell_m^2 \sin^2 \Omega t \, dt$$

$$\overline{\ell^2} = \frac{\ell_m^2}{2} \tag{4.7}$$

and the *root mean square value (rms value)* is

$$\ell_{rms} = \sqrt{\overline{\ell^2}} = \frac{\ell_m}{\sqrt{2}} \approx 0.707 \, \ell_m \tag{4.8}$$

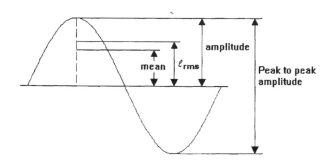

Figure 4.2. *Characteristics of a single sinusoid*

Thus

$$\ell_{\text{rms}} = \frac{\pi}{2\sqrt{2}} \, \bar{\ell} \qquad\qquad [4.9]$$

this can then be written in the more general form [BRO]:

$$\ell_{\text{rms}} = F_f \, \bar{\ell} = \frac{1}{F_c} \, \ell_m \qquad\qquad [4.10]$$

The F_f and F_c factors are, respectively, termed the *form factor and peak factor*. These parameters give, in real cases where the signal is not pure, some indications on its form and its resemblance to a sinusoid. For a pure sinusoid:

$$F_f = \frac{\pi}{2\sqrt{2}} \approx 1.11 \qquad\qquad [4.11]$$

and

$$F_c = \sqrt{2} \approx 1.414 \qquad\qquad [4.12]$$

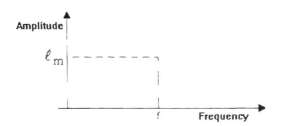

Figure 4.3. *Spectrum of a sinusoid (line spectrum)*

Such a signal is also termed *simple harmonic*. Its spectrum comprises of only one line at a particular frequency.

The spectrum of a signal made up of several sinusoids is known as *discrete* (spectrum *of lines)* [BEN 71].

NOTE: *The mean square value is. generally. a total measurement of the static and dynamic components of the vibratory signal. The continuous component can be separated by calculating the mean value [BEN 63] [PEN 65]:*

$$\bar{\ell} = \frac{1}{T} \int_0^T \ell(t) \, dt$$

This is zero for a perfect sinusoid. the time axis being centred, while the dynamic part is characterised by calculating the centred mean square value (variance).

$$s_\ell^2 = \frac{1}{T} \int_0^T \left[\ell(t) - \bar{\ell} \right]^2 \, dt \tag{4.13}$$

We then have

$$\overline{\ell^2} = s_\ell^2 + \left(\bar{\ell} \right)^2 \tag{4.14}$$

The variance is equal to the mean square value if $\bar{\ell} = 0$.

4.1.4. *Periodic excitation*

Movements encountered in the real environment are seldom purely sinusoidal. Some are simply periodic, the signal being repeated are regular time intervals T_1. With rare exceptions, a periodic signal can be represented by a Fourier series, i.e. by a sum of purely sinusoidal signals:

$$\ell(t) = \frac{a_0}{2} + \sum_{n=1}^{\infty} \left(a_n \cos 2 \pi n f_1 t + b_n \sin 2 \pi n f_1 t \right) \tag{4.15}$$

where

$$f_1 = \frac{1}{T} = \text{fundamental frequency}$$

$$a_n = \frac{2}{T} \int_0^T \ell(t) \cos 2 \pi n f t \, dt$$

$$b_n = \frac{2}{T} \int_0^T \ell(t) \sin 2 \pi n f t \, dt$$

$$(n = 0, 1, 2, 3 ...)$$

All the frequencies $f_n = n\,f_1$ are multiple integers of the fundamental frequency f_1.

For the majority of practical applications, it is sufficient to know the amplitude and the frequency of the various components, the phase being neglected. The representation of such a periodic signal can then be made as in Figure 4.4, by a discrete spectrum giving the amplitude ℓ_{m_n} of each component according to its frequency.

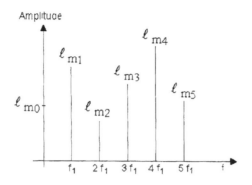

Figure 4.4. *Spectrum of a periodic signal*

With each component being sinusoidal, the rms value $\ell_{rms_n} = \dfrac{\ell_{m_n}}{\sqrt{2}}$ or the mean value of $\left| \ell_n(t) \right|$, $\ell_{n\,mean} = \dfrac{2}{\pi} \ell_{m_n}$ can easily be drawn against f. These various parameters give information on the excitation severity, but are alone insufficient to describe it since they do not give any idea of the frequency. $\ell(t)$ can be also written [PEN 65]:

$$\ell(t) = \ell_{m_0} + \sum_{n=1}^{\infty} \ell_{m_n} \sin\left(2\,\pi\,n\,f_1\,t - \varphi_n\right) \qquad [4.16]$$

where

ℓ_{m_n} = amplitude of the nth component

ϕ_n = phase of the nth component

L_0 = continuous component

$$\ell_n(t) = \ell_{mn} \sin\left(2\,\pi\,n\,f_1 - \varphi_n\right)$$

$$\ell_{m0} = \frac{a_0}{2}$$

$$\ell_{mn} = \sqrt{a_n^2 + b_n^2} \quad (n = 1, 2, 3...)$$

$$\varphi = \text{Arc tan}\ \frac{b_n}{a_n}$$

The periodic signal $\ell(t)$ can thus be regarded as the sum of a constant component and an infinite number (or not) of sinusoidal components, named *harmonics,* whose frequencies are multiple integers of f.

The Fourier series can be characterized entirely by the coefficients a_n and b_n at the frequencies n f_1 and can be represented by line spectra giving a_n and b_n versus the frequency. If we do not consider the phases φ_n as is often the case in practice, we can also draw a line spectrum giving the coefficients ℓ_{mn} versus the frequency.

The ordinate axis can indicate the amplitude of each component or its rms value. We have [FOU 64]:

$$\bar{\ell} = \ell_{m0} \tag{4.17}$$

$$\ell(t) = \sum_{n=1}^{\infty} \ell_{mn} \sin\left(2\,\pi\,f_n\,t + \varphi_n\right)$$

$$\ell_{rms}^2 = \frac{1}{T}\sum \int_0^T \ell_{mn}^2 \sin\left(2\,\pi\,n\,f_1\,t + \varphi_p\right) dt$$

$$+ \frac{2}{T}\sum \int_0^T \ell_{mp}\,\ell_{mq}\,\sin\left(2\,\pi\,p\,f_1\,t + \varphi_p\right)\sin\left(2\,\pi\,q\,f_1\,t + \varphi_q\right) dt$$

$$+ \frac{1}{T}\sum \int_0^T \ell_{mn}^2 \sin^2\left(2\,\pi\,n\,f_1\,t + \varphi_n\right) dt$$

$$= \frac{1}{T}\sum \ell_{mn}^2 \int_0^T \frac{1}{2}\left\{1 - \cos\left[2\left(2\,\pi\,n\,f_1\,t + \varphi_n\right)\right]\right\}\,dt = \frac{1}{T}\sum \ell_{mn}^2 \int_0^T \frac{dt}{2}$$

The second term, the integral over one period of the product of two sinusoidal functions, is zero. If the mean value is zero

$$\ell_{rms}^2 = \frac{1}{T} \sum_{n=1}^{\infty} \int_0^T \ell_{mn}^2 \frac{dt}{2} = \sum_{n=1}^{\infty} \frac{\ell_{mn}^2}{2} \qquad [4.18]$$

Each component has as a mean square value equal to

$$\overline{\ell_n^2} = \frac{1}{2} \ell_{mn}^2 \qquad [4.19]$$

If the mean value is not zero

$$\overline{\ell^2} = \ell_0^2 + \frac{1}{2} \sum_{n=1}^{\infty} \ell_{mn}^2 \qquad [4.20]$$

the variance is given by

$$s_\ell^2 = \overline{\ell^2} - \left(\overline{\ell}\right)^2 = \frac{1}{2} \sum_{n=1}^{\infty} \ell_{mn}^2 \qquad [4.21]$$

The relations [4.2] giving $\dot{\ell}(t)$ and $\ddot{\ell}(t)$ from $\ell(t)$ do not directly apply any more (it is necessary to derive each term from the sum). The forms of each one of these curves are different.

The mean value and the rms value of $\ell(t)$ can always be calculated from the general expressions [BRO] [KLE 71b].

4.1.5. *Quasi periodic signals*

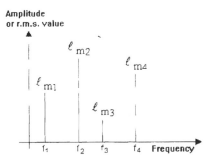

Figure 4.5. *Spectrum of a quasi periodic signal*

A signal made up of the sum of several periodic signals will not in itself periodic be if all the possible ratios between the frequencies of the components are irrational numbers; the resulting signal can then be written

$$\ell(t) = \sum_{n=1}^{\infty} \ell_{m_n} \sin\left(2 \pi f_n t + \varphi_n\right) \qquad [4.22]$$

If we also neglect here also the phases φ_n we can still represent $\ell(t)$ graphically by a line spectrum.

4.2. Periodic and sinusoidal vibrations in the real environment

Perfectly sinusoidal vibrations are seldom met in the real environment. In certain cases, however, the signal can be treated similarly to a sinusoid in order to facilitate the analyses. Such vibrations are observed, for example, in rotating machines, and in badly balanced parts in rotation (inbalances of the trees, defects of coaxiality of the reducer shafts with the driving shafts, electric motor, gears) [RUB 64].

The more frequent case of periodic vibrations decomposable in Fourier series is reduced to a sinusoidal vibrations problem, by studying the effect of each harmonic component and by applying the superposition theorem (if the necessary assumptions are respected, in particular that of linearity). They can be observed on machines generating periodic impacts (presses), in internal combustion engines with several cylinders and so on [BEN 71], [BRO], [KLE 71b] and [TUS 72].

Quasi-periodic vibrations can be studied in the same manner, component by component, insofar as each component can be characterized. They are measured, for example, in plane structures propelled by several badly synchronized engines [BEN 71].

4.3. Sinusoidal vibration tests

The sinusoidal vibration tests carried out using electrodynamic shakers or hydraulic vibration machines can have several aims:

– the simulation of an environment of the same nature;

– the search for resonance frequencies (identification of the dynamic behaviour of a structure). This research can be carried out by measuring the response of the structure at various points when it is subjected to random excitation, shocks or *swept frequency* sinusoidal vibrations. In this last case, the frequency of the sinusoid varies over time according to a law which is in general exponential, sometimes linear. When the swept sine test is controlled by an analogical control system, the frequency varies in a continuous way with time. When numerical control means are used, the frequency remains constant at a certain time with each selected value, and varies between two successive values by increments that may or may not be constant depending on the type of sweeping selected;

Example

The amplitude is supposed to be x_m = 10 cm at a frequency of 0.5 Hz.

Maximum velocity:

$$\dot{x}_m = 2 \pi f x_m = 0.314 \text{ m/s}$$

Maximum acceleration:

$$\ddot{x}_m = (2 \pi f)^2 x_m = 0.987 \text{ m/s}^2$$

At 3 Hz, x_m = 10 cm

Velocity:

$$\dot{x}_m = 1.885 \text{ m/s}$$

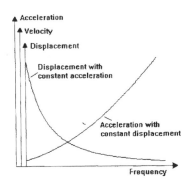

Figure 4.6. *Acceleration, velocity and displacement versus frequency*

Acceleration:

$$\ddot{x}_m = 35.53 \ m/s^2$$

At 10 Hz, if $\ddot{x}_m = 5 \ m/s^2$, the velocity is equal to $\dot{x}_m = \dfrac{\ddot{x}_m}{2\pi f} = 0.0796 \ m/s$ and

the displacement is $x_m = \dfrac{\ddot{x}_m}{(2\pi f)^2} = 1.27 \ 10^{-3} \ m.$

– fatigue tests either on test-bars or directly on structures, the frequency of the sinusoid often being fixed on the resonance frequency of the structure. In this last case, the test is often intended to simulate the fatigue effects of a more complex real environment, generally random and making the assumption that induced fatigue is dominant around resonance [GAM 92]. The problems to be solved are then the following [CUR 71]:

– the determination of an equivalence between random and sinusoidal vibration. There are rules to choose the severity and the duration of an *equivalent sine test* [GAM 92];

– it is necessary to know the frequencies of resonance of the material (determined by a preliminary test);

– for these frequencies, it is necessary to choose the number of test frequencies, in general lower than the number of resonances (in order for a sufficient fraction of the total time of test to remain at each frequency), and then to define the severity, and the duration of each sinusoid at each frequency of resonance selected. The choice of the frequencies is very important. As far as possible, those for which rupture by fatigue is most probable are chosen i.e. those for which the Q factor is higher than a given value (2generally). This choice can be questioned since, being based on previously measured transfer functions, it is a function of the position of the sensors and can thus lead to errors;

– the frequent control of the resonance frequency, which varies appreciably at the end of the material's lifetime.

For the sine tests, the specifications specify the frequency of the sinusoid, its duration of application and its amplitude.

The amplitude of the excitation is in general defined by a zero to peak acceleration (sometimes peak-to-peak); for very low frequencies (less than a few Hertz), it is often preferred to describe the excitation by a displacement because the acceleration is in general very weak. With intermediate frequencies, velocity is sometimes chosen.

Chapter 5

Response of a linear single-degree-of-freedom mechanical system to a sinusoidal excitation

In Chapter 2 simple harmonic movements, damped and undamped, were considered, where the mechanical system, displaced from its equilibrium position and released at the initial moment, was simply subjected to a restoring force and, possibly, to a damping force.

In this chapter the movement of a system subjected to steady state excitation, whose amplitude varies sinusoidally with time, with its restoring force in the same direction will be studied. The two possibilities of an excitation defined by a force applied to the mass of the system or by a movement of the support of the system, this movement itself being defined by an displacement, a velocity or an acceleration varying with time will also be examined.

The two types of excitation focussed on will be:

– the case close to reality where there are damping forces:

– the ideal case where damping is zero.

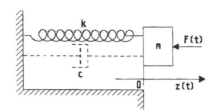

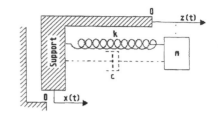

Figure 5.1. *Excitation by a force*   **Figure 5.2.** *Excitation by an acceleration*

5.1. General equations of motion

5.1.1. Relative response

In Chapter 2 the differential equation of the movement was established. The Laplace transform is written, for the relative response:

$$Q(p) = \frac{\Lambda(p)}{p^2 + 2\,\xi\,p + 1} + \frac{p\,q_0 + (\dot{q}_0 + 2\,\xi\,q_0)}{p^2 + 2\,\xi\,p + 1} \qquad [5.1]$$

q_0 and $\dot{q}_0$ being initial conditions. To simplify the calculations, and by taking account of the remarks of this chapter, it is supposed that $q_0 = \dot{q}_0 = 0$. If this were not the case, it would be enough to add to the final expression of $q(\theta)$ the term $C(\theta)$ previously calculated.

The transform of a sinusoid

$$\lambda(\theta) = \sin h\,\theta \qquad [5.2]$$

is given by

$$\Lambda(p) = \frac{h}{p^2 + h^2} \qquad [5.3]$$

where $h = \dfrac{\Omega}{\omega_0}$ (Ω being the pulsation of the sinusoid and ω_0 the natural pulsation of the undamped one-degree-of-freedom mechanical system), yielding

$$Q(p) = \frac{h}{\left(p^2 + h^2\right)\left(p^2 + 2\,\xi\,p + 1\right)} \qquad [5.4]$$

Case 1: $0 \le \xi \le 1$ (*underdamped* system)

$$Q(p) = \frac{-h}{\left(1-h^2\right)^2 + 4\,\xi^2\,h^2} \left[\frac{2\,\xi\,p + h^2 - 1}{p^2 + h^2} - \frac{2\,\xi\,p + 4\,\xi^2 + h^2 - 1}{p^2 + 2\,\xi\,p + 1}\right] \qquad [5.5]$$

$$q(\theta) = \frac{\left(1-h^2\right)\sin h\theta - 2\,\xi\,h\cos h\theta}{\left(1-h^2\right)^2 + 4\,\xi^2\,h^2} + h e^{-\xi\theta}\,\frac{2\xi\cos\sqrt{1-\xi^2}\,\theta + \dfrac{2\xi^2 + h^2 - 1}{\sqrt{1-\xi^2}}\sin\sqrt{1-\xi^2}\,\theta}{\left(1-h^2\right)^2 + 4\,\xi^2\,h^2}$$

$$[5.6]$$

For non-zero initial conditions, this must be added to $q(\theta)$

$$C(\theta) = e^{-\xi\theta}\left[q_0\,\cos\sqrt{1-\xi^2}\,\theta + \frac{\dot{q}_0 + q_0\,\xi}{\sqrt{1-\xi^2}}\sin\sqrt{1-\xi^2}\,\theta\right] \qquad [5.7]$$

Case 2: $\xi = 1$ (*critical damping*)

For zero initial conditions,

$$Q(p) = \frac{h}{\left(p^2 + h^2\right)(p+1)^2} \qquad [5.8]$$

$$q(\theta) = L^{-1}\left\{\frac{-h}{\left(1+h^2\right)^2}\left[\frac{2\,p + h^2 - 1}{p^2 + h^2} - \frac{2\,p + h^2 + 3}{(p+1)^2}\right]\right\} \qquad [5.9]$$

$$q(\theta) = -\frac{h}{\left(1+h^2\right)^2}\left\{2\cos h\,\theta + \frac{h^2 - 1}{h}\sin h\,\theta - e^{-\theta}\left(2 + \theta + h^2\,\theta\right)\right\} \qquad [5.10]$$

For non-zero initial conditions, we have to add to $q(\theta)$:

$$C(\theta) = \left[q_0 + \left(q_0 + \dot{q}_0\right)\theta\right]e^{-\theta} \qquad [5.11]$$

Case 3: $\xi > 1$ (*overdamped* system)

$$Q(p) = \frac{-h}{\left(1-h^2\right)^2 + 4\,\xi^2\,h^2}\left[\frac{2\,\xi\,p + h^2 - 1}{p^2 + h^2} - \frac{2\,\xi\,p + 4\,\xi^2 + h^2 - 1}{p^2 + 2\,\xi\,p + 1}\right] \qquad [5.12]$$

The denominator can be written $p^2 + 2\,\xi\,p + 1$, for $\xi > 1$,

$$p^2 + 2\,\xi\,p + 1 = \left(p + \xi + \sqrt{\xi^2 - 1}\right)\left(p + \xi - \sqrt{\xi^2 - 1}\right) \qquad [5.13]$$

yielding

$$q(\theta) = \frac{-h}{\left(1-h^2\right)^2 + 4\,\xi^2\,h^2}\left\{2\,\xi\cos h\,\theta + \frac{h^2 - 1}{h}\sin h\,\theta\right.$$

$$+2\xi\,\frac{\left(\xi - \sqrt{\xi^2 - 1}\right)e^{-\left(\xi - \sqrt{\xi^2-1}\right)\theta} - \left(\xi + \sqrt{\xi^2 - 1}\right)e^{-\left(\xi + \sqrt{\xi^2-1}\right)\theta}}{2\sqrt{\xi^2 - 1}}$$

$$\left. +\left(4\xi^2 + h^2 - 1\right)\frac{e^{-\left(\xi + \sqrt{\xi^2-1}\right)\theta} - e^{-\left(\xi - \sqrt{\xi^2-1}\right)\theta}}{2\sqrt{\xi^2 - 1}}\right\}$$

$$\boxed{q(\theta) = \frac{\left(1-h^2\right)\sinh\theta - 2\xi h\cosh\theta}{\left(1-h^2\right)^2 + 4\xi^2 h^2} + h\,e^{-\xi\theta}\,\frac{2\xi\,\mathrm{ch}\sqrt{\xi^2 - 1}\,\theta + \dfrac{h^2 + 2\xi^2 - 1}{\sqrt{\xi^2 - 1}}\,\mathrm{sh}\sqrt{\xi^2 - 1}\,\theta}{\left(1-h^2\right)^2 + 4\xi^2 h^2}}$$

$$[5.14]$$

with, for non-zero initial conditions,

$$C(\theta) = \frac{e^{-\xi\theta}}{\sqrt{\xi^2 - 1}}\left[\left(\xi\,q_0 + \dot{q}_0\right)\mathrm{sh}\sqrt{\xi^2 - 1}\,\theta + q_0\sqrt{\xi^2 - 1}\,\mathrm{ch}\sqrt{\xi^2 - 1}\,\theta\right] \qquad [5.15]$$

5.1.2. *Absolute response*

Case 1: $0 \leq \xi < 1$

Zero initial conditions

$$Q(p) = \frac{h \left(1 + 2 \xi p\right)}{\left(p^2 + h^2\right)\left(p^2 + 2 \xi p + 1\right)} \tag{5.16}$$

$$Q(p) = \frac{h}{\left(1 - h^2\right)^2 + 4 \xi^2 h^2} \left\{ \frac{2 \xi h^2 p + h^2 - 1}{p^2 + 2 \xi p + 1} + \frac{-2 \xi h^2 p + 4 \xi^2 h^2 + 1 - h^2}{p^2 + h^2} \right\} \tag{5.17}$$

$$q(\theta) = \frac{\left(1 - h^2 + 4 \xi^2 h^2\right) \sin h\,\theta - 2 \xi h^3 \cos h\,\theta}{\left(1 - h^2\right)^2 + 4 \xi^2 h^2}$$

$$- h\, e^{-\xi \theta} \frac{\dfrac{1 - h^2 + 2 \xi^2 h^2}{\sqrt{1 - \xi^2}} \sin \sqrt{1 - \xi^2}\,\theta - 2 \xi h^2 \cos \sqrt{1 - \xi^2}\,\theta}{\left(1 - h^2\right)^2 + 4 \xi^2 h^2} \tag{5.18}$$

If the initial conditions are not zero, it must be added to $q(\theta)$

$$C(\theta) = e^{-\xi \theta} \left[q_0 \cos \sqrt{1 - \xi^2}\,\theta + \frac{\dot{q}_0 + \xi\left(q_0 - 2 \lambda_0\right)}{\sqrt{1 - \xi^2}} \sin \sqrt{1 - \xi^2}\,\theta \right] \tag{5.19}$$

Case 2: $\xi = 1$

$$Q(p) = \frac{h \left(1 + 2 p\right)}{\left(p^2 + h^2\right)\left(p + 1\right)^2} \tag{5.20}$$

$$Q(p) = \frac{h}{\left(1 + h^2\right)^2} \left[\frac{2 h^2}{p + 1} - \frac{1 + h^2}{\left(p + 1\right)^2} - \frac{2 h^2 p}{p^2 + h^2} + \frac{3 h^2 + 1}{p^2 + h^2} \right] \tag{5.21}$$

$$q(\theta) = \frac{1}{\left(1+h^2\right)^2}\left\{h\left[2\,h^2-\left(1+h^2\right)\theta\right]e^{-\theta}+\left(3\,h^2+1\right)\sin h\,\theta-2\,h^3\cos h\,\theta\right\}$$

[5.22]

Non-zero initial conditions

$$C(\theta) = \left[q_0+\left(q_0+\dot{q}_0-2\,\lambda_0\right)\theta\right]e^{-\theta}$$

[5.23]

Case 3: $\xi > 1$

Zero initial conditions

$$Q(p) = \frac{h\left(1+2\,\xi\,p\right)}{\left(p^2+h^2\right)\left(p^2+2\,\xi\,p+1\right)}$$

[5.24]

$$Q(p) = \frac{h}{\left(1-h^2\right)^2+4\,\xi^2 h^2}\left\{\frac{2\,\xi\,h^2 p+h^2-1}{p^2+2\,\xi\,p+1}+\frac{-2\,\xi\,h^2 p+4\,\xi^2 h^2+1-h^2}{p^2+h^2}\right\}$$

[5.25]

i.e.

$$q(\theta) = \frac{h\,e^{-\xi\theta}}{\left(1-h^2\right)^2+4\,\xi^2 h^2}\left\{\frac{h^2-1-2\,\xi^2 h^2}{\sqrt{\xi^2-1}}\,sh\sqrt{\xi^2-1}\,\theta+2\,\xi\,h^2 ch\sqrt{\xi^2-1}\,\theta\right\}$$

$$+\frac{\left(4\,\xi^2\,h^2+1-h^2\right)\sin h\,\theta-2\,\xi\,h^3\cos h\,\theta}{\left(1-h^2\right)^2+4\,\xi^2\,h^2}$$

[5.26]

Non-zero initial conditions

$$C(\theta) = \frac{e^{-\xi\theta}}{\sqrt{\xi^2-1}}\left\{\left[\xi\left(q_0-2\,\lambda_0\right)+\dot{q}_0\right]sh\sqrt{\xi^2-1}\,\theta+\sqrt{\xi^2-1}\,q_0\,ch\sqrt{\xi^2-1}\,\theta\right\}$$

[5.27]

This must be added to $q(\theta)$.

5.1.3. *Summary*

The principal relations obtained for zero initial conditions are brought together below.

Relative response

$0 \leq \xi < 1$

$$q(\theta) = \frac{\left(1 - h^2\right) \sin h\,\theta - 2\,\xi\,h\,\cos h\,\theta}{\left(1 - h^2\right)^2 + 4\,\xi^2\,h^2}$$

$$+ h\,e^{-\xi\,\theta}\,\frac{2\xi\cos\sqrt{1 - \xi^2}\;\theta + \dfrac{h^2 + 2\xi^2 - 1}{\sqrt{1 - \xi^2}}\,\sin\sqrt{1 - \xi^2}\;\theta}{\left(1 - h^2\right)^2 + 4\xi^2 h^2}$$

$\xi = 1$

$$q(\theta) = \frac{h}{\left(1 + h^2\right)^2}\left\{\frac{1 - h^2}{h}\,\sin h\,\theta - 2\cos h\,\theta + \left(2 + \theta + h^2\theta\right)e^{-\theta}\right\}$$

$\xi > 1$

$$q(\theta) = \frac{\left(1 - h^2\right)\sin h\theta - 2\xi h\,\cos h\theta}{\left(1 - h^2\right)^2 + 4\xi^2 h^2}$$

$$+ h\,e^{-\xi\,\theta}\,\frac{2\,\xi\,ch\sqrt{\xi^2 - 1}\,\theta + \dfrac{h^2 + 2\xi^2 - 1}{\sqrt{\xi^2 - 1}}\,sh\sqrt{\xi^2 - 1}\,\theta}{\left(1 - h^2\right)^2 + 4\,\xi^2\,h^2}$$

Absolute response

$0 \leq \xi < 1$

$$q(\theta) = \frac{\left(1 - h^2 + 4\,\xi^2 h^2\right)\sin h\,\theta - 2\,\xi\,h^3\cos h\,\theta}{\left(1 - h^2\right)^2 + 4\,\xi^2\,h^2}$$

$$- h\,e^{-\xi\,\theta}\,\frac{\dfrac{1 - h^2 + 2\,\xi^2\,h^2}{\sqrt{1 - \xi^2}}\,\sin\sqrt{1 - \xi^2}\;\theta - 2\,\xi\,h^2\cos\sqrt{1 - \xi^2}\;\theta}{\left(1 - h^2\right)^2 + 4\,\xi^2\,h^2}$$

$\xi = 1$

$$q(\theta) = \frac{1}{\left(1 + h^2\right)^2}\left\{h\left[2\,h^2 - \left(1 + h^2\right)\theta\right]e^{-\theta} + \left(3\,h^2 + 1\right)\sin h\theta - 2\,h^3 \cos h\theta\right\}$$

$\xi > 1$

$$q(\theta) = \frac{\left(1 - h^2\right)\sin h\theta - 2\xi h \cos h\theta}{\left(1 - h^2\right)^2 + 4\xi^2 h^2}$$

$$+ h\,e^{-\xi\theta}\;\frac{2\xi ch\sqrt{\xi^2 - 1}\,\theta + \dfrac{h^2 + 2\xi^2 - 1}{\sqrt{\xi^2 - 1}}\,sh\sqrt{\xi^2 - 1}\,\theta}{\left(1 - h^2\right)^2 + 4\xi^2 h^2}$$

5.1.4. Discussion

Whatever the value of ξ, the response $q(\theta)$ is made up of three terms:

– the first, $C(\theta)$, related to conditions initially non-zero, which disappears when θ increases, because of the presence of the term $e^{-\xi\theta}$;

– the second, which corresponds to the transient movement at the reduced frequency $\sqrt{1 - \xi^2}$ resulting from the sinusoid application at time $\theta = 0$. This oscillation attenuates and disappears after a while from ξ because of the factor $e^{-\xi\theta}$. In the case of the relative response, for example, for $0 \le \xi < 1$, this term is equal to

$$h\,e^{-\xi\theta}\;\frac{2\,\xi\cos\sqrt{1 - \xi^2}\,\theta + \dfrac{h^2 + 2\,\xi^2 - 1}{\sqrt{1 - \xi^2}}\sin\sqrt{1 - \xi^2}\,\theta}{\left(1 - h^2\right)^2 + 4\,\xi^2\,h^2}$$

– the third term corresponds to an oscillation of reduced pulsation h, which is that of the sinusoid applied to the system. The vibration of the system *is forced*, the frequency of the response being imposed on the system by the excitation. The sinusoid applied theoretically having one unlimited duration, it is said that the response, described by this third term, is *steady state*.

Table 5.1. *Expressions for reduced response*

Response → Excitation ↓ (Base movement)	Displacement Absolute $y(t)$	Displacement Relative $z(t)$	Velocity Absolute $\dot{y}(t)$	Velocity Relative $\dot{z}(t)$	Acceleration Absolute $\ddot{y}(t)$	Acceleration Relative $\ddot{z}(t)$	Reaction force on base $F_T(t)$
Displacement $x(t)$	$\dfrac{y}{x_m}$	$\dfrac{z}{h^2\,x_m}$	$\dfrac{\dot{y}}{h\,\omega_0\,x_m}$	$\dfrac{\dot{z}}{h^2\,\omega_0\,x_m}$	$\dfrac{\ddot{y}}{h^2\,\omega_0^2\,x_m}$	$\dfrac{\ddot{z}}{h^4\,\omega_0^2\,x_m}$	
Velocity $\dot{x}(t)$	$\dfrac{h\,\omega_0\,y}{\dot{x}_m}$	$\dfrac{\omega_0\,z}{h\,\dot{x}_m}$	$\dfrac{\dot{y}}{\dot{x}_m}$	$\dfrac{\dot{z}}{h^2\,\dot{x}_m}$	$\dfrac{h\,\omega_0\,\dot{y}}{\dot{x}_m}$	$\dfrac{\ddot{z}}{h^3\,\omega_0\,\dot{x}_m}$	
Acceleration $\ddot{x}(t)$	$\dfrac{h^2\,\omega_0^2\,y}{\ddot{x}_m}$	$\dfrac{\omega_0^2\,z}{\ddot{x}_m}$	$\dfrac{h\,\omega_0\,\dot{y}}{\ddot{x}_m}$	$\dfrac{\omega_0\,\dot{z}}{h\,\ddot{x}_m}$	$\dfrac{\ddot{y}}{\ddot{x}_m}$	$\dfrac{\ddot{z}}{h^2\,\ddot{x}_m}$	
Force on the mass m (here, $z \equiv y$)	$\dfrac{k\,z}{F_m}$		$\dfrac{\sqrt{k\,m}\;\dot{z}}{h\,F_m}$		$\dfrac{m\,\ddot{z}}{h^2\,F_m}$		$\dfrac{F_T}{F_m}$

The steady state response for $0 \leq \xi < 1$ will be considered in detail in the following paragraphs. The reduced parameter $q(\theta)$ is used to calculate the response of the mechanical system. This is particularly interesting because of the possibility of easily deducing expressions for relative or absolute response $q(\theta)$, irrespective of the way the excitation (force, acceleration, velocity or displacement of the support) is defined, as in Table 5.1.

5.1.5. Response to periodic excitation

The response to a periodic excitation can be calculated by development of a Fourier series for the excitation [HAB 68].

$$\ell(t) = \frac{a_0}{2} + \sum_{n=1}^{\infty} \left(a_n \cos n \, \Omega \, t + b_n \sin n \, \Omega \, t \right) \tag{5.28}$$

$$a_0 = \frac{2}{T} \int_0^T \ell(t) \, dt \tag{5.29}$$

$$a_n = \frac{2}{T} \int_0^T \ell(t) \cos n \, \Omega \, t \quad dt \tag{5.30}$$

$$b_n = \frac{2}{T} \int_0^T \ell(t) \sin n \, \Omega \, t \quad dt \tag{5.31}$$

The response of a one-degree-of-freedom system obeys the differential equation

$$\ddot{u}(t) + 2 \, \xi \, \omega_0 \, \dot{u}(t) + \omega_0^2 \, u(t) = \omega_0^2 \, \ell(t) \tag{5.32}$$

$$\ddot{u}(t) + 2 \, \xi \, \omega_0 \, \dot{u}(t) + \omega_0^2 \, u(t) = \omega_0^2 \left[\frac{a_0}{2} + \sum_{n=1}^{\infty} \left(a_n \cos n \, \Omega \, t + b_n \sin n \, \Omega \, t \right) \right] \tag{5.33}$$

This equation being linear, the solutions of the equation calculated successively for each term in sine and cosine can be superimposed. This yields

$$u(t) = \frac{a_0}{2} + \sum_{n=1}^{\infty} \frac{a_n \cos\left(n \, \Omega \, t - \phi_n \right) + b_n \sin\left(n \, \Omega \, t - \phi_n \right)}{\sqrt{\left(1 - n^2 \frac{\Omega^2}{\omega_0^2} \right)^2 + \left(n \frac{\Omega}{\omega_0} \right)^2}} \tag{5.34}$$

with

$$\phi_n = \text{Arc tan} \; \frac{n \; \dfrac{\Omega}{\omega_0}}{1 - n^2 \; \dfrac{\Omega^2}{\omega_0^2}}$$

[5.35]

5.1.6. *Application to calculation for vehicle suspension response*

Considering a vehicle rolling at velocity v on a sinusoidal road as shown in Figure 5.3.

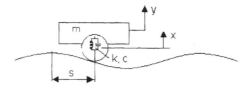

Figure 5.3. *Example of a vehicle on road*

$$x = X \cos \frac{2 \pi s}{L}$$

[5.36]

s = distance between a maximum of sinusoid and the vehicle

L = sinusoid.period

It is supposed that [VOL 65]:

– the wheels are small, so that the hub of each wheel is at a constant distance from the road;

– the tires have negligible deformation.

We have with the notation already used in the preceding paragraphs:

$$m \; \ddot{y} + c \left(\dot{y} - \dot{x} \right) + k \left(y - x \right) = 0$$

[5.37]

$$m \; \ddot{y} + c \; \dot{y} + k \; y = k \; x + c \; \dot{x}$$

[5.38]

$$\ddot{y} + 2\,\xi\,\omega_0\,\dot{y} + \omega_0^2\,y = \omega_0^2\,x + 2\,\xi\,\omega_0\,x \qquad [5.39]$$

The distance s is related to time by $s = v\,t$ yielding

$$x = X\,\cos\Omega\,t \qquad\qquad [5.40]$$

with

$$\Omega = \frac{2\,\pi\,v}{L} \qquad\qquad [5.41]$$

$$\ddot{y} + 2\,\xi\,\omega_0\,\dot{y} + \omega_0^2\,y = \omega_0^2\,X\,\cos\Omega\,t - 2\,\xi\,\omega_0\,\Omega\,\sin\Omega\,t \qquad [5.42]$$

$$\ddot{y} + 2\,\xi\,\omega_0\,\dot{y} + \omega_0^2\,y = \omega_0^2\,X\,\sqrt{1 + \left(2\,\xi\,h\right)^2}\;\cos\!\left(\Omega\,t + \theta\right) \qquad [5.43]$$

where

$$\tan\theta = 2\,\xi\,h \qquad\qquad [5.44]$$

$$h = \frac{\Omega}{\omega_0}$$

$$y = x\,\cos\!\left(\Omega\,t + \theta - \varphi\right) \qquad\qquad [5.45]$$

$$\boxed{\,y = x\,\sqrt{\dfrac{1 + \left(2\,\xi\,h\right)^2}{\left(1 - h^2\right)^2 + \left(2\,\xi\,h\right)^2}}\,} \qquad [5.46]$$

$$\tan\varphi = \frac{2\,\xi\,h}{1 - h^2} \qquad\qquad [5.47]$$

The displacement y must be smallest possible to make the suspension effective. It is necessary therefore that h or the velocity be large. If ξ tends towards zero, y tends towards infinity when h tends towards 1, with critical velocity

$$v_{cr} = \frac{\omega_0\,L}{2\,\pi} \qquad\qquad [5.48]$$

When ξ is different from zero, the value of y for $h = 1$ is

$$y = x \sqrt{1 + \frac{1}{(2\,\xi)^2}}$$

[5.49]

5.2. Transient response

5.2.1. *Relative response*

For $0 \le \xi < 1$

The response

$$q(\theta) = h \ e^{-\xi\,\theta} \ \frac{2\,\xi\,\cos\sqrt{1-\xi^2}\ \theta - \dfrac{h^2 + 2\,\xi^2 - 1}{\sqrt{1-\xi^2}}\sin\sqrt{1-\xi^2}\ \theta}{\left(1-h^2\right)^2 + 4\,\xi^2\,h^2}$$

[5.50]

can be also written

$$q(\theta) = e^{-\xi\,\theta}\ A(h)\ \sin\!\left(\sqrt{1-\xi^2}\ \theta - \alpha\right)$$

[5.51]

where

$$A(h) = \frac{h}{\sqrt{1-\xi^2}\ \sqrt{\left(1-h^2\right)^2 + 4\,\xi^2\,h^2}}$$

[5.52]

and

$$\tan\,\alpha = \frac{2\,\xi\,\sqrt{1-\xi^2}}{1-h^2 - 2\,\xi^2}$$

[5.53]

A *pseudo-sinusoidal movement* occurs. The total response $q(\theta)$ is zero for $\theta = 0$ since the term representing the transient response is then equal to

$$q_T(0) = \frac{2\,\xi\,h}{\left(1-h^2\right)^2 + 4\,\xi^2\,h^2}$$

[5.54]

This response q_T never takes place alone. It is superimposed on the steady state response $q_P(\theta)$ studied in the following paragraph.

The amplitude $A(h)$ is maximum when $\dfrac{dA(h)}{dh} = 0$, i.e. when

$$\frac{dA(h)}{dh} = \frac{1}{\sqrt{1-\xi^2}} \frac{1-h^4}{\left[\left(1-h^2\right)^2 + 4\,\xi^2\,h^2\right]^{3/2}} = 0 \qquad [5.55]$$

$\dfrac{dA(h)}{dh} = 0$ when $h = 1$ $(h \geq 0)$.

In this case,

$$A_m(h) = \frac{1}{2\,\xi\,\sqrt{1-\xi^2}} \qquad [5.56]$$

The movement has a logarithmic decrement equal to [KIM 29]:

$$\delta = \frac{2\,\pi\,\xi}{\sqrt{1-\xi^2}} \qquad [5.57]$$

and for the reduced pseudo-period $\dfrac{2\,\pi}{\sqrt{1-\xi^2}}$.

The transient response q_T has an amplitude equal to $\dfrac{1}{N}^{th}$ of the first peak after cycle number n such that $\dfrac{2\,\pi\,\xi}{\sqrt{1-\xi^2}} = \dfrac{1}{n}\ln N$

i.e.

$$n = \frac{\sqrt{1-\xi^2}}{2\,\pi\,\xi}\ln N \qquad [5.58]$$

For ξ small, it becomes

$$n \approx \frac{\ln N}{2\,\pi\,\xi} = \frac{Q}{\pi}\ln N$$

i.e.

$$n \approx \frac{Q \ln N}{\pi}$$ [5.59]

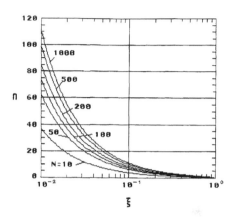

Figure 5.4. *Cycle number for attenuation* N *of the transient relative response*

If N ≈ 23, we have n ≈ Q. When a system is subjected to a sinewave excitation, the amplitude of the response is established gradually during the transitional stage up to a level proportional to that of the excitation and which corresponds to the steady state response. In the following paragraph it is seen that, if $h = \sqrt{1 - \xi^2}$, the response tends in steady state mode towards

$$H_m = \frac{1}{2 \xi \sqrt{1 - \xi^2}} .$$

The number of cycles necessary to reach this steady state response is independent of h. For ξ small, this number is roughly proportional to the Q factor of the system.

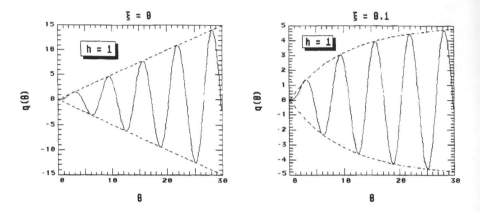

Figure 5.5. *Establishment of the relative response for* $\xi = 0$

Figure 5.6. *Establishment of the relative response for* $\xi = 0.1$

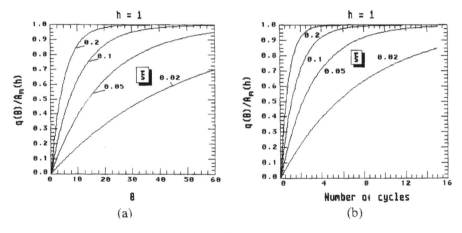

Figure 5.7. *Ratio of transient response/steady state response*

For the particular case where $\xi = 1$

$$q_T(\theta) = \frac{h}{\left(1 + h^2\right)^2} \left(2 + \theta + h^2\, \theta\right) e^{-\theta}$$

[5.60]

5.2.2. *Absolute response*

For $0 \leq \xi < 1$

$$q(\theta) = \frac{2\,\xi\,h^2\,\cos\sqrt{1-\xi^2}\;\theta + \dfrac{h^2 - 1 - 2\,\xi^2\,h^2}{\sqrt{1-\xi^2}}\,\sin\sqrt{1-\xi^2}\;\theta}{\left(1 - h^2\right)^2 + 4\,\xi^2\,h^2}\; h\,e^{-\xi\theta} \qquad [5.61]$$

or

$$q(\theta) = e^{-\xi\theta}\,B(h)\,\sin\left(\sqrt{1-\xi^2}\;\theta - \beta\right) \qquad\qquad [5.62]$$

with

$$B(h) = \frac{h}{\sqrt{1-\xi^2}\,\sqrt{\left(1-h^2\right)^2 + 4\,\xi^2\,h^2}} = A(h) \qquad\qquad [5.63]$$

and

$$\tan\beta = \frac{2\,\xi\,\sqrt{1-\xi^2}\;h^2}{h^2 - 1 - 2\,\xi^2\,h^2} \qquad\qquad [5.64]$$

If $\xi = 1$

$$q_T(\theta) = \frac{h}{\left(1+h^2\right)^2}\left[2\,h^2 - \left(1+h^2\right)\theta\right]e^{-\theta} \qquad\qquad [5.65]$$

5.3. Steady state response

5.3.1. *Relative response*

For $0 \leq \xi < 1$, the steady state response is written

$$q(\theta) = \frac{\left(1 - h^2\right)\sin h\,\theta - 2\,\xi\,h\,\cos h\,\theta}{\left(1 - h^2\right)^2 + 4\,\xi^2\,h^2} \qquad\qquad [5.66]$$

This expression can be also put in the form

$$q(\theta) = H(h) \sin(h\,\theta - \varphi) \tag{5.67}$$

In the amplitude of this response H_{RD} it is noted that, the first index recalls that the response is relative and the second is about a displacement. Therefore

$$H(h) = \frac{1}{\sqrt{\left(1 - h^2\right)^2 + 4\,\xi^2\,h^2}} = H_{RD}(h) \tag{5.68}$$

The phase is such that

$$\tan \varphi = \frac{2\,\xi\,h}{1 - h^2} \tag{5.69}$$

5.3.2. Absolute response

For $0 \le \xi < 1$, the steady state response is expressed

$$q(\theta) = \frac{\left(1 - h^2 + 4\,\xi^2\,h^2\right)\sin h\,\theta - 2\,\xi\,h^3\,\cos h\,\theta}{\left(1 - h^2\right)^2 + 4\,\xi^2\,h^2} \tag{5.70}$$

As previously, this response can be written

$$q(\theta) = \frac{\sqrt{1 + 4\,\xi^2\,h^2}\,\sin(h\,\theta - \varphi)}{\sqrt{\left(1 - h^2\right)^2 + 4\,\xi^2\,h^2}} = H_{AD}\,\sin(h\,\theta - \varphi) \tag{5.71}$$

where

$$H_{AD} = \sqrt{\frac{1 + 4\,\xi^2\,h^2}{\left(1 - h^2\right)^2 + 4\,\xi^2\,h^2}} \tag{5.72}$$

and

$$\tan \varphi = \frac{2\,\xi\,h^3}{1 - h^2 + 4\,\xi^2\,h^2} \tag{5.73}$$

H_{AD} is termed the *transmissibility factor* or *transmissibility* or *transmittance*.

5.4. Responses $\left|\dfrac{\omega_0 \, \dot{z}}{\ddot{x}_m}\right|$, $\left|\dfrac{\omega_0 \, z}{\dot{x}_m}\right|$ and $\dfrac{\sqrt{k \, m} \, \dot{z}}{F_m}$

Starting with the study of the responses $\left|\dfrac{\omega_0 \, \dot{z}}{\ddot{x}_m}\right|$, $\left|\dfrac{\omega_0 \, z}{\dot{x}_m}\right|$ and $\dfrac{\sqrt{k \, m} \, \dot{z}}{F_m}$, some important definitions are introduced. They are equal to

$$\dot{q}(\theta) = \frac{h}{\sqrt{\left(1 - h^2\right)^2 + 4 \, \xi^2 \, h^2}} \; \sin\left(h \, \theta - \psi\right) = H_{RV} \, \sin\left(h \, \theta - \psi\right) \qquad [5.74]$$

where

$$H_{RV} = \frac{h}{\sqrt{\left(1 - h^2\right)^2 + 4 \, \xi^2 \, h^2}} \qquad [5.75]$$

More interesting is the case where the input is an acceleration $\ddot{x}_m$, and the reduced response $q(\theta)$ gives the relative displacement $z(t)$ yielding

$$\left|\dot{q}(\theta)\right| = \left|\frac{\omega_0 \, \dot{z}}{\ddot{x}_m}\right| = H_{RD} \, h \, \cos\left(h \, \theta - \psi\right) \qquad [5.76]$$

$$\left|\dot{q}(\theta)\right| = H_{RV} \, h \, \sin\left(h \, \theta - \psi\right) \qquad [5.77]$$

with

$$H_{RV} = h \, H_{RD} \qquad [5.78]$$

and

$$\psi = \varphi - \frac{\pi}{2} \qquad [5.79]$$

5.4.1. *Variations of velocity amplitude*

5.4.1.1. *Quality factor*

The amplitude H_{RV} of the velocity passes through a maximum when the derivative $\dfrac{dH_{RV}}{dh}$ is zero.

$$\frac{dH_{RV}}{dh} = \frac{1 - h^4}{\left[\left(1 - h^2\right)^2 + 4\,\xi^2\,h^2\right]^{3/2}} \qquad [5.80]$$

This function is equal to zero when $h = 1$ ($h \geq 0$). The response thus is at a maximum (whatever be ξ) for $h = 1$. There is then *velocity resonance*, and

$$H_{RV_{max}} = \frac{1}{2\,\xi} = Q \qquad [5.81]$$

At resonance, the amplitude of the forced vibration $\dot{q}(\theta)$ is Q times that of the excitation (here the physical significance of the Q factor is seen). It should be noted that this resonance takes place for a frequency equal to the natural frequency of the undamped system, and not for a frequency equal to that of the free oscillation of the damped system. It tends towards the unit when h tends towards zero. The curve thus starts from the origin with a slope equal to 1 (whatever ξ). For $h = 0$, $H_{RV} = 0$.

The slope tends towards zero when $h \to \infty$, like H_{RV}. The expression of H_{RV} does not change when h is replaced by $\dfrac{1}{h}$; thus taking a logarithmic scale for the abscissae, the curves $H_{RV}(h)$ are symmetrical with respect to the line $h = 1$.

For $\xi = 0$,

$$H_{RV} = \frac{h}{\left|1 - h^2\right|} \qquad [5.82]$$

$H_{RV} \to \infty$ when $h \to 1$.

Since $\tan \psi = \tan\left(\varphi - \dfrac{\pi}{2}\right) = \cotan \varphi = \dfrac{h^2 - 1}{2\,\xi\,h}$,

H_{RV} can be written in the form

$$H_{RV} = \frac{1}{2\,\xi\,\sqrt{1 + \tan^2 \psi}}$$ [5.83]

Setting $y = 2\,\xi\,H_{RV}$ and $x = \tan\psi$, the curves

$$\begin{cases} y = \dfrac{1}{\sqrt{1 + x^2}} \\ \psi = \text{Arc}\tan x \end{cases}$$

valid for all the systems m, k, and c are known as *universal.*

In the case of an excitation by force, the quantity $2\,\xi\,H_{RV}$ is equal to

$$2\,\frac{c}{2\,\sqrt{k\,m}}\,\frac{\sqrt{k\,m}\,\dot{z}}{F_m} = \frac{c\,\dot{z}}{F_m}$$

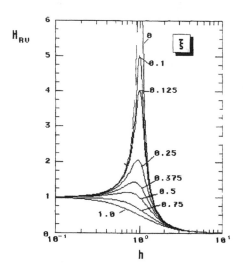

Figure 5.8. *Amplitude of velocity response*

5.4.1.2. *Hysteresis loop*

It has been supposed up to now that damping was viscous, with the damping force being proportional to the relative velocity between the ends of the damping device and of the form $F_d = -c\,\dot{z}$, where c is the damping constant, and in a direction opposed to that of the movement. This damping, which leads to a linear differential equation of the movement, is itself known as *linear*. If the relative displacement response $z(t)$ has the form

$$z = z_m\,\sin(\Omega\,t - \varphi)$$

the damping force is equal to

$$F_d(t) = c\,z_m\,\Omega\,\cos(\Omega\,t - \varphi) = F_{d_m}\,\cos(\Omega\,t - \varphi) \qquad [5.84]$$

where

$$F_{d_m} = c\,\Omega\,z_m \qquad [5.85]$$

The curve $F_d(z)$ (hysteresis loop) has the equations, in parametric coordinates,

$$\begin{cases} z = z_m\,\sin(\Omega\,t - \varphi) \\ F_d = F_{d_m}\,\cos(\Omega\,t - \varphi) \end{cases}$$

i.e., after elimination of time t:

$$\frac{F_d^2}{F_{d_m}^2} + \frac{z^2}{z_m^2} = 1 \qquad [5.86]$$

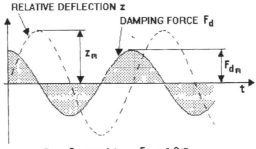

$$F_d = F_{d_m}\cos\Omega\,t \; ; \quad F_{d_m} = c\,\Omega\,z_m$$

Figure 5.9. *Viscous damping force*

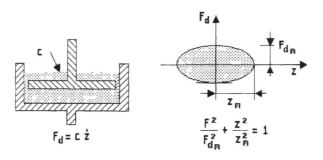

Figure 5.10. *Hysteresis loop for viscous damping [RUZ 71]*

The hysteresis loop is thus presented in the form of an ellipse of half the smaller axis $F_{d_m} = c \, \Omega \, z_m$ and half the larger axis z_m.

5.4.1.3. *Energy dissipated during a cycle*

The energy dissipated during a cycle can be written:

$$\Delta E_d = \int_{1 \text{ cycle}} |F_d| \; dz = \int_0^T |F_d| \; \frac{dz}{dt} dt$$

$$\Delta E_d = c \int_0^T \dot{z}^2 \; dt$$

Knowing that $z(t) = z_m \sin(\Omega t - \varphi)$, we have

$$\Delta E_d = c \, z_m^2 \, \Omega^2 \int_0^{2\pi/\Omega} \cos^2(\Omega t - \varphi) \, dt$$

i.e., since $\cos^2(\Omega t - \varphi) = \dfrac{1 + \cos[2 \, (\Omega t - \varphi)]}{2}$

$$\Delta E_d = \pi \, c \, \Omega \, z_m^2 \tag{5.87}$$

or [CRE 65]:

$$\Delta E_d = \pi \, z_m \, F_{d_m} \tag{5.88}$$

For a viscously damped system, in which the damping constant c is independent of the frequency, the relative damping ξ is inversely proportional to the frequency:

$$\xi = \frac{c}{2\sqrt{k\,m}} = \frac{c}{2\pi\,m\,f_0} \tag{5.89}$$

We can deduce from this the energy Δ consumed per time unit. If T is the period of the excitation $\left(T = \dfrac{2\pi}{\Omega} \right)$,

$$\Delta = \frac{\Delta E_d}{T} = \frac{\pi\,c\,\Omega}{T}\,z_m^2 = \frac{1}{2}\,c\Omega^2\,z_m^2 = \frac{1}{2}\,c\,\dot{z}_m^2$$

$$\Delta = \frac{1}{2}\,c\,\Omega^2\,z_m^2 = \xi\,\omega_0\,m\,\Omega^2\,z_m^2 \tag{5.90}$$

Since (Chapter 3) $z_m = \dfrac{z_s}{\sqrt{\left(1-h^2\right)^2 + 4\,\xi^2\,h^2}}$.

$$\Delta = \xi\,\omega_0\,m\,\Omega^2\,\frac{z_s^2}{\left(1-h^2\right)^2 + 4\,\xi^2\,h^2}$$

$$\Delta = \xi\,\omega_0^3\,m\,z_s^2\,\frac{h^2}{\left(1-h^2\right)^2 + 4\,\xi^2\,h^2} \tag{5.91}$$

Consumed energy is at a maximum when the function $\dfrac{h^2}{\left(1-h^2\right)^2 + 4\,\xi^2\,h^2}$, equal to $H_{RV}^2(h)$, is at a maximum, i.e. for $h = 1$, yielding

$$\Delta_m = \frac{\omega_0^2\,m\,z_s^2}{4\,\xi} \tag{5.92}$$

and

$$\frac{\Delta}{\Delta_m} = \frac{4\,\xi^2\,h^2}{\left(1-h^2\right)^2 + 4\,\xi^2\,h^2} = H_{RV}^2(h) \tag{5.93}$$

The dissipated energy is thus inversely proportional to ξ. When ξ decreases, the resonance curve $\Delta(h)$ presents a larger and narrower peak [LAN]. It can however be shown that the surface under the curve $\Delta(h)$ remains unchanged.

This surface S is described by:

$$S = \int_0^\infty \Delta(\Omega)\, d\Omega = \int_0^\infty \xi\, \omega_0^3\, m\, z_s^2\, \frac{h^2}{\left(1 - h^2\right)^2 + 4\,\xi^2\, h^2}\, d\Omega \qquad [5.94]$$

$$S = \xi\, \omega_0^4\, m\, z_s^2 \int_0^\infty \frac{h^2}{\left(1 - h^2\right)^2 + 4\,\xi^2\, h^2}\, dh$$

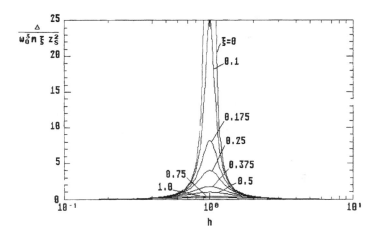

Figure 5.11. *Energy dissipated by damping*

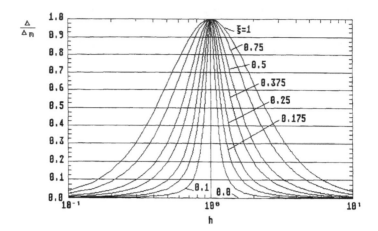

Figure 5.12. *Reduced energy dissipated by damping*

The integral is equal to $\dfrac{\pi}{4\,\xi}$ (Volume 4), yielding

$$S = \pi\, m\, \omega_0^4\, z_s^2 = \pi\, \omega_0^2\, k\, z_s^2 \tag{5.95}$$

The surface S is thus quite independent of ξ. Therefore

$$\frac{S}{\Delta_m} = \pi\, \xi\, \omega_0 \tag{5.96}$$

5.4.1.4. *Half-power points*

The *half-power points* are defined by the values of h such as the energy dissipated per unit time is equal to $\dfrac{\Delta_m}{2}$ yielding

$$\frac{1}{2}\,\frac{\omega_0^2\, m\, z_s^2}{4\,\xi} = \xi\, \omega_0^3\, m\, z_s^2 \, \frac{h^2}{\left(1-h^2\right)^2 + 4\,\xi^2\, h^2}$$

$$\left(1-h^2\right)^2 + 4\,\xi^2\, h^2 = 8\,\xi^2\, h^2$$

$$\frac{h^2 - 1}{2 \xi h} = \pm 1$$

i.e., since h and ξ are positive,

$$\begin{cases} h_1 = -\xi + \sqrt{1 + \xi^2} \\ h_2 = +\xi + \sqrt{1 + \xi^2} \end{cases}$$ [5.97]

A logarithmic scale is sometimes used to represent the transmissibility and a unit, the *Bel*, is introduced, or generally in practice, a subunit, *the decibel*. It is said that a power P_1 is higher by n decibels (dB) than a power P_0 if

$$10 \log \frac{P_1}{P_0} = n$$ [5.98]

If $P_1 > P_0$, the system has a gain of n dB. If $P_1 < P_0$, the system produces an attenuation of n dB [GUI 63]. If instead of the powers, forces or velocities are considered here, the definition of the gain (or attenuation which is none other than a negative gain) remains identical on the condition of replacing the constant 10 by factor of 20 ($\log P = 2 \log V_e + \text{Constant}$), since the power is proportional to the square of the rms velocity [LAL 95a].

The curve $2 \xi H_{RV}$ or H_{RV} is close to a horizontal line for ξ small, passes through a maximum for $h = 1$, then decreases and tends towards zero when h becomes large compared to 1. By analogy with an electric resonant system, the mechanical system can be characterized by the interval (bandwidth) between two frequencies h_1 and h_2 selected in such a way that $2 \xi H_{RV}$ is either equal to $\frac{1}{\sqrt{2}}$, or for h such that

$$H_{RV}(h) = \frac{Q}{\sqrt{2}}$$ [5.99]

The values h_1 and h_2 correspond to the abscissae of two points N_1 and N_2 named *half-power points* because the power which can be dissipated by the shock absorber during a simple harmonic movement at a given frequency is proportional to the square of the reduced amplitude H_{RV} [MEI 67].

5.4.1.5. *Bandwidth*

If $2 \, \xi \, H_{RV} = \dfrac{1}{\sqrt{2}}$, $\sqrt{1 + \tan^2 \psi} = \sqrt{2}$, i.e. $\tan^2 \psi = 1$, yielding

$$\frac{h^2 - 1}{2\xi h} = Q\left(h - \frac{1}{h} \right) = \pm 1 \qquad\qquad [5.100]$$

The quantity $Q\left(h - \dfrac{1}{h} \right)$ is *the dissonance*. It is zero with resonance and equivalent to $Q(h-1)$ in its vicinity [GUI 63]. The condition $\tan \psi = \pm 1$, is $\psi = \pm \dfrac{\pi}{4}$ (modulo π) which shows that ψ undergoes, when h varies from h_1 to h_2, a variation of $-\dfrac{\pi}{4}$ to $\dfrac{\pi}{4}$. i.e. of $\dfrac{\pi}{2}$.

h_1 and h_2 are calculated from [5.100]

$$\left\{ \begin{array}{l} Q\left(h_2 - \dfrac{1}{h_2} \right) = 1 \\[3mm] Q\left(h_1 - \dfrac{1}{h_1} \right) = -1 \end{array} \right. \qquad\qquad [5.101]$$

This becomes

$$\left\{ \begin{array}{l} (h_2 - h_1)\left(1 + \dfrac{1}{h_1 \, h_1} \right) = \dfrac{2}{Q} \\[3mm] h_1 \, h_2 = \dfrac{1}{Q} \end{array} \right. \qquad\qquad [5.102]$$

$$h_2 - h_1 = \frac{1}{Q} \qquad\qquad [5.103]$$

The bandwidth of the system $\Delta h = h_2 - h_1$ can be also written

$$\boxed{\Omega_2 - \Omega_1 = \frac{\omega_0}{Q}} \qquad\qquad [5.104]$$

This is all the narrower when Q is larger.

Selectivity

In the general case of a system having its largest response for a pulsation ω_m, the *selectivity* σ is defined by

$$\sigma = \frac{\omega_m}{\Delta\Omega}(= \frac{h_m}{\Delta h}) \qquad\qquad [5.105]$$

where $\Delta\Omega$ is the previously definite bandwidth. σ characterizes the function of the filter of the system, by its ability, by eliminating near-frequencies (of $\Delta\Omega$) to allow through a single frequency. For a resonant system, $\omega_m = \omega_0$ and $\sigma = Q$.

In electricity, the interval $\Omega_2 - \Omega_1$ characterizes the selectivity of the system. The resonant system is a simple model of a filter where the selective transmissibility can make it possible to choose signals in the useful band (Ω_1, Ω_2) among other signals external with this band and which are undesirable. The selectivity is improved as the peak becomes more acute. In mechanics, this property is used for protection against vibrations (filtering by choosing the frequency of resonance smaller than the frequency of the vibration).

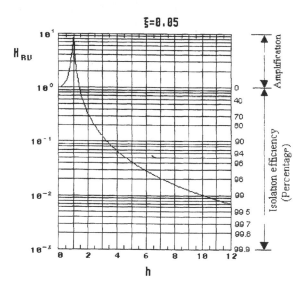

Figure 5.13. *Domains of the transfer function* H_{RV}

It can also be shown [LAL 95a] [LAL 95b] that the response of a one-degree-of-freedom system is primarily produced by the contents of the excitation in this frequency band.

From these relations, the expressions of h_1 and h_2 can be deduced:

$$h_1 = -\xi + \sqrt{1 + \xi^2} \ (\approx 1 - \xi \ \text{if} \ \xi \ \text{is small})$$

[5.106]

$$h_2 = +\xi + \sqrt{1 + \xi^2} \ (\approx 1 + \xi \ \text{if} \ \xi \ \text{is small})$$

[5.107]

The bandwidth $\Delta h = h_2 - h_1$ can be also written

$$\Delta h = 2 \xi = \frac{1}{Q}$$

[5.108]

yielding, since $h = \dfrac{\Omega}{\omega_0}$,

$$\boxed{Q = \frac{\omega_0}{\Delta \Omega} = \frac{f_0}{\Delta f}}$$

[5.109]

NOTE: *The ratio* $\dfrac{\Omega - \omega_0}{\omega_0} = h - 1$ *is also sometimes considered. For the abscissae* Ω_1 *and* Ω_2 *of the half-power points, and for* ξ *small, this ratio is equal, respectively, to* $-\dfrac{1}{2 Q}$ *and* $+\dfrac{1}{2Q}$.

The Q factor of the mechanical systems does not exceed a few tens of units and those of the electric circuits do not exceed a few hundreds.

In [2.139] it was seen that

$$\delta \approx 2 \pi \xi$$

yielding [GUR 59]:

$$\boxed{\delta \approx \frac{\pi}{Q}}$$

[5.110]

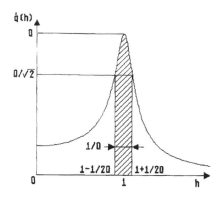

Figure 5.14. *Bandwidth*

The bandwidth can thus be also defined as the field of the frequencies transmitted with an attenuation of $10 \log 2 \approx 3.03$ dB below the maximum level (attenuation between the levels Q and $\dfrac{Q}{\sqrt{2}}$) [DEN 56], [THU 71].

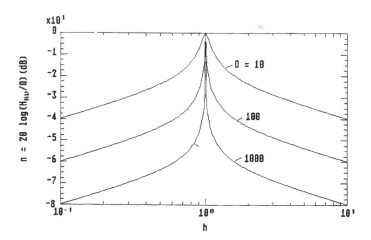

Figure 5.15. *Representation in dB of the transfer function* H_{RV}

Figure 5.15 represents some resonance curves, plotted versus the variable h and for various values of the Q factor with the ordinate being in dB [GUI 63].

5.4.2. *Variations in velocity phase*

In [5.77] it was seen that

$$\dot{q}(\theta) = H_{RV} \sin(h\,\theta - \psi)$$

where [5.79]

$$\psi = \varphi - \frac{\pi}{2}$$

yielding

$$\tan \psi = -\frac{1}{\tan \varphi} = \frac{h^2 - 1}{2\,\xi\,h} \qquad\qquad [5.111]$$

To obtain the curves $\psi(h)$, it is therefore enough to shift by $\dfrac{\pi}{2}$ the already plotted curves $\varphi(h)$, while keeping ξ the same. The phase ψ varies from $-\dfrac{\pi}{2}$ to $+\dfrac{\pi}{2}$ since φ varies from 0 to π. It is zero for $h = 1$, i.e. when the frequency of the system is equal to that of the excitation (whatever ξ).

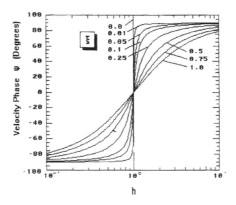

Figure 5.16. *Velocity phase versus* h

The velocity of the mass is thus always in phase with the excitation in this case.

When h is lower than 1, the velocity of the mass is in advance of the excitation ($\psi < 0$, i.e. $-\psi > 0$). When h is larger than 1, the velocity of the mass has a phase lag with respect to excitation. In passing through resonance, the curve $\psi(h)$ presents a point of inflection. Around this point there is then a roughly linear variation of the phase varying with h (in an interval that is larger as ξ becomes smaller).

5.5. Responses $\dfrac{k\,z}{F_m}$ and $\dfrac{\omega_0^2\,z}{\ddot{x}_m}$

In these cases,

$$q(\theta) = H_{RD}(h)\ \sin\left(h\,\theta - \varphi\right) \qquad\qquad [5.112]$$

The response $q(\theta)$ is at a maximum when $\sin(h\,\theta - \varphi) = 1$. i.e. for

$$h\,\theta - \varphi = \left(4\,k + 1\right)\frac{\pi}{2}.$$

5.5.1. *Variation in response amplitude*

5.5.1.1. *Dynamic amplification factor*

Given that the excitation is a force applied to the mass or an acceleration communicated to the support, the reduced response makes it possible to calculate the relative displacement z. The ratio H_{RD} of the amplitude of the response relative to displacement to the equivalent static displacement ($\dfrac{F_m}{k}$ or $\dfrac{\ddot{x}_m}{\omega_0^2}$) is often called the

dynamic amplification factor

NOTE: *Some authors [RUZ 71] call the amplification factor of the quantities* $\dfrac{k\,z}{F_m}$,

$\dfrac{\sqrt{k\,m}\,\dot{z}}{F_m}$ *or* $\dfrac{m\,\ddot{z}}{F_m}$ *(amplification factor of the displacement, of the velocity and of*

the acceleration respectively) and relative transmissibility $\dfrac{\ddot{z}}{\ddot{x}_m}$, $\dfrac{\dot{z}}{\dot{x}_m}$ *or* $\dfrac{z}{x_m}$

(acceleration, velocity or displacement).

The function $H_{RD}(h)$ depends on the parameter ξ. It is always a positive function which passes through a maximum when the denominator passes through a minimum. The derivative of $\left(1 - h^2\right)^2 + 4\,\xi^2\,h^2$ is cancelled when

$$h_m = \sqrt{1 - 2\,\xi^2}$$ [5.113]

($h \geq 0$), provided that $1 - 2\xi^2 \geq 0$, i.e. $\xi \leq \dfrac{1}{\sqrt{2}}$. When h tends towards zero, $H_{RD}(h)$ tends towards 1 whatever the value of ξ. There is resonance for $h = h_m$, the function $H_{RD}(h)$ is maximum and is then equal to

$$H_m = \frac{1}{2\,\xi\,\sqrt{1 - \xi^2}}$$ [5.114]

When $h \to \infty$, $H_{RD}(h) \to 0$. In addition, $H_m \to \infty$ when $\xi \to 0$. In this case, $h_m = 1$. Resonance is all the more acute since the relative damping ξ is smaller; the damping has two effects: it lowers the maximum and makes the peak less acute.

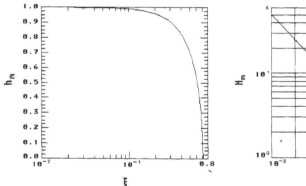

Figure 5.17. *Frequency of the maximum of* H_{RD} *versus* ξ

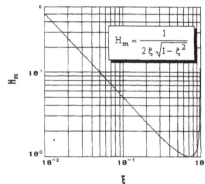

Figure 5.18. *Maximum of* H_{RD} *versus* ξ

It can be interesting to chart H_m versus h; it can be seen that the calculation of ξ versus h_m from [5.113] gives $\xi = \sqrt{\dfrac{1 - h_m^2}{2}}$.

This yields:

$$H_m = \frac{1}{\sqrt{1 - h_m^4}}$$ [5.115]

where h_m can only be positive. Here interest will focus just on the branch of the curve belonging to the interval $0 \leq h \leq 1$.

There can be a maximum only for $h \leq 1$ (i.e. for a frequency of the excitation lower than that of the resonator ω_0), the condition $\xi \leq \dfrac{1}{\sqrt{2}}$, is assumed to be realized.

If $h = 1$ is not a condition of resonance, then there is resonance only if at the same time $\xi = 0$. Otherwise, resonance takes place when $h < 1$.

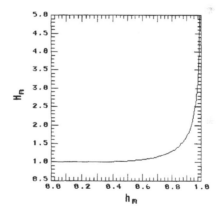

Figure 5.19. *Maximum of* H_{RD} *versus the peak frequency*

It can be seen that the condition $\xi = 1$ corresponds to the *critical modes.*

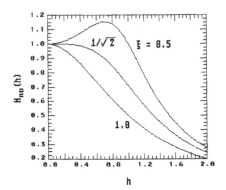

Figure 5.20. *Dynamic amplification factor around the critical mode*

Like all the curves of $H(h)$, the one corresponding to $\xi = \dfrac{1}{\sqrt{2}}$, which separates the areas from the curves with or without a maximum, has a horizontal level in the vicinity of the ordinate axis ($h = 0$).

$\xi = \dfrac{1}{\sqrt{2}}$ gives optimum damping. It is for this value that H_m varies less versus h (an interesting property in electro-acoustics).

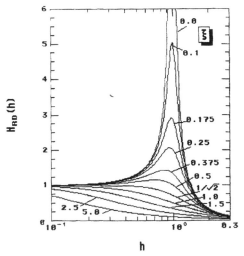

Figure 5.21. *Dynamic amplification factor for various values of* ξ

It can also be seen that when $\xi = \dfrac{1}{\sqrt{2}}$, the three first derivatives from H_m are zero for $h = 0$.

Finally it should be noted that this $\xi = \dfrac{1}{\sqrt{2}}$ value is lower than that for critical damping ($\xi = 1$). It could be thought that the existence of the transient state ($\xi < 1$) does not disturb the response but in practice it has little influence. Setting δ, the logarithmic decrement, it was shown that

$$\delta = \frac{2 \pi \xi}{\sqrt{1 - \xi^2}}$$

For $\xi = \dfrac{1}{\sqrt{2}}$, thus $\delta = 2 \pi$. This is an enormous damping: the ratio of two successive maximum displacements is then equal to $e^{\delta} = e^{2\pi} \approx 560$. The transient state disappears very quickly and is negligible as of the second oscillation.

5.5.1.2. *Width of* $H(h)$ *for* $H_{RD} = \dfrac{H_{RD\,max}}{\sqrt{2}}$

By analogy with the definition of the half-power points given for H_{RV} in paragraph 5.4.1.4. we can calculate the width Δh of the peak of H_{RD} for the ordinate $H_{RD} = \dfrac{H_{RD\,max}}{\sqrt{2}}$. It has been seen that $H_{RD\,max} = \dfrac{Q}{\sqrt{1 - \xi^2}}$ yielding

$$H_{RD} \equiv \frac{1}{\sqrt{\left(1 - h^2\right)^2 + \dfrac{h^2}{Q^2}}} = \frac{Q}{\sqrt{2}\,\sqrt{1 - \xi^2}} \qquad [5.116]$$

and

$$h^2 = 1 - \frac{1}{2\,Q^2} \pm \frac{1}{Q} \sqrt{1 - \frac{1}{4\,Q^2}} \qquad (Q \geq \frac{1}{2}, \text{ i.e. } \xi \leq 1)$$

$$h^2 = 1 - 2\,\xi^2 \pm 2\,\xi \sqrt{1 - \xi^2}$$

h^2 must be positive, which requires for the first root that $1 + 2\xi\sqrt{1-\xi^2} \ge 2\xi^2$. The other root leads to $2\xi^2 + 2\xi\sqrt{1-\xi^2} \le 1$. Let us set h_1 and h_2 the two roots.

This gives

$$h_2^2 - h_1^2 = 1 - 2\xi^2 + 2\xi\sqrt{1-\xi^2} - 1 + 2\xi^2 + 2\xi\sqrt{1-\xi^2}$$

$$h_2^2 - h_1^2 = 4\xi\sqrt{1-\xi^2} \qquad\qquad [5.117]$$

If ξ is small, $h^2 \approx 1 \pm 2\xi$, $h \approx \sqrt{1 \pm 2\xi} \approx 1 \pm \xi$

$$h_2^2 - h_1^2 \approx 4\xi$$

$$h_2 - h_1 \approx 2\xi \text{ and } h_2 + h_1 \approx 2$$

Particular case

If ξ is small with respect to 1, we have, at first approximation,

$$h \approx \sqrt{1 \pm 2\xi} \ (h \ge 0)$$

$$h \approx 1 \pm \xi \qquad\qquad [5.118]$$

In the particular case where ξ is small, the abscissa of the points for which $H_{RD} = \dfrac{H_{RD\,max}}{\sqrt{2}}$ is approximately equal to the abscissa of the half-power points (definite from H_{RV}). The bandwidth can be calculated from

$$\Delta h = h_2 - h_1 \qquad\qquad [5.119]$$

5.5.2. Variations in phase

The phase is given by

$$\tan\varphi = \frac{2\xi h}{1 - h^2} \qquad\qquad [5.120]$$

It is noted that:

– $\left|\tan \varphi\right|$ is unchanged when h is replaced by $\dfrac{1}{h}$;

– $\tan \varphi \to \infty$ when $h \to 1$, therefore $\varphi \to \dfrac{\pi}{2}$: the response is in quadrature phase lead with respect to the excitation;

. – $\tan \varphi = 0$, i.e. $\varphi = 0$ when $h = 0$ (in the interval considered);

– the derivatives below do not cancel;

$$\frac{d\varphi}{dh} = \frac{2\,\xi\left(1 + h^2\right)}{\left(1 - h^2\right)^2 + 4\,\xi^2\,h^2}$$

[5.121]

– $\tan \varphi \to 0$, i.e. $\varphi \to \pi$, when $h \to \infty$ (φ cannot tend towards zero since there is no maximum. The function which is cancelled already when $h = 0$ cannot cancel one second time): the response and the excitation are in opposite phase;

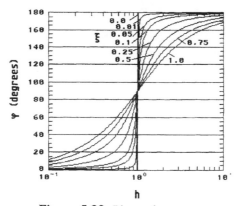

Figure 5.22. *Phase of response*

– for all values of ξ, φ is equal to $\dfrac{\pi}{2}$ when $h = 1$; all the curves thus pass through the point $h = 1$, $\varphi = \dfrac{\pi}{2}$;

–for $\xi < 1$, all the curves have a point of inflection in $h = 1$, $\varphi = \dfrac{\pi}{2}$. The slope at this point is greater as ξ gets smaller.

Particular cases

– For $h = \sqrt{1 - 2\xi^2}$ (resonance) and $\xi \le \dfrac{1}{\sqrt{2}}$

$$\tan \varphi = \frac{2\,\xi\,h}{1 - h^2} = \frac{\sqrt{1 - 2\xi^2}}{\xi} \qquad\qquad [5.122]$$

$$\varphi = \text{Arc}\tan \frac{\sqrt{1 - 2\xi^2}}{\xi} \qquad\qquad [5.123]$$

– When h is small, the mass m practically has a movement in phase with the excitation ($\varphi \approx 0$). In this case, q_{max} being closer to 1 as h is smaller, the mass follows the movement of the support.

Values of angle φ ranging between 180° and 360° cannot exist because, in this case, the shock absorber would provide energy to the system instead of dissipating it [RUB 64].

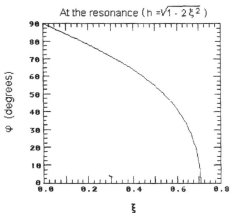

Figure 5.23. *Resonance phase*

Since $H_{RD} \approx 1$ for h small, for an excitation by force on the mass, $\dfrac{k\,z}{F_m} \approx 1$, i.e.

$z \approx \dfrac{F_m}{k}$. The response is controlled in a dominating way by the stiffness of the system. In this domain where h is small with respect to the unit, the dimensioning calculations of structure in statics can be carried out by taking the values of H_{RD} at the frequency of the vibration, in order to take account at the same time of the static load and of the small dynamic amplification. These calculations can possibly be supplemented by a fatigue analysis if this phenomenon is considered to be important [HAL 75].

– for h = 1, the maximum value of q(θ) is

$$q_{max} = \frac{1}{2 \, \xi \, \sqrt{1 - \xi^2}} \approx Q \qquad\qquad [5.124]$$

and the phase is

$$\varphi \rightarrow + \frac{\pi}{2} \qquad\qquad [5.125]$$

$$q(\theta) = \frac{\sin\left(h \, \theta + \frac{\pi}{2} \right)}{2 \, \xi \, \sqrt{1 - \xi^2}} \qquad\qquad [5.126]$$

$$q(\theta) = \frac{\cos h \, \theta}{2 \, \xi \, \sqrt{1 - \xi^2}} \qquad\qquad [5.127]$$

The amplitude of the response is a function of the damping ξ. It is larger if ξ is smaller. The movement is out of phase by $\dfrac{\pi}{2}$ with respect to the excitation.

If the excitation is a force, at the resonance, $H_{RD} = \dfrac{1}{2 \, \xi \, \sqrt{1 - \xi^2}}$, i.e.

$$z_m = \frac{-F_m}{2 \, k \, \xi \, \sqrt{1 - \xi^2}} \qquad\qquad [5.128]$$

$$z_m \approx \frac{-F_m}{2 \, k \, \xi} = \frac{-F_m}{c \, \omega_0} \qquad\qquad [5.129]$$

Here, analysis must be of the dynamic type, the response being able to be equal to several times the equivalent static excitation.

– for h >> 1,

$$q(\theta) \approx \frac{\sin(h\ \theta - \varphi)}{h^2}$$ [5.130]

where $\varphi = -\pi$:

$$q(\theta) \approx -\frac{\sin h\ \theta}{h^2}$$ [5.131]

If the excitation is a force, we have

$$H_{RD} \approx \frac{1}{h^2}$$ [5.132]

i.e.

$$z_m \approx \frac{F_m}{k\ h^2}$$ [5.133]

$$z_m \approx \frac{F_m}{m\ \Omega^2}$$ [5.134]

where Ω = pulsation of the excitation.

The response is primarily a function of the mass m. It is smaller than the equivalent static excitation.

According to whether h satisfies one or the other of these three conditions, one of the three elements stiffness, damping or mass thus has a dominating action on the resulting movement of the system [BLA 61] [RUB 64].

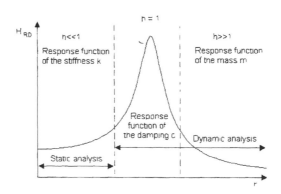

Figure 5.24. *Fields of dynamic amplification factor*

Particular case where $\xi = 0$

$$q(\theta) = \frac{\sin\left(h\,\theta - \varphi\right)}{1 - h^2} \qquad\qquad\qquad [5.135]$$

(The positive root for $h < 1$ is chosen in order to preserve at $q(\theta)$ the same sign for $\xi = 0$ rather than for ξ which is very small in the expression [5.66]).

$$q_{max} = H_{RD} = \frac{1}{1 - h^2} \qquad\qquad\qquad [5.136]$$

The variations of q_{max} versus h are represented in Figure 5.25. It is noted that, when h tends towards 1, q_{max} tends towards infinity. It is necessary here to return to the assumptions made and to remember that the system is considered linear, which supposes that the amplitude of the variations of the response q remains small. This curve $q_{max}(h)$ thus does not make sense in the vicinity of the asymptote.

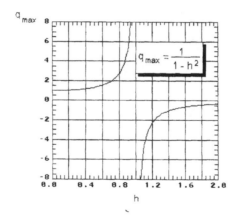

Figure 5.25. *Variations of* q_{max} *versus* h

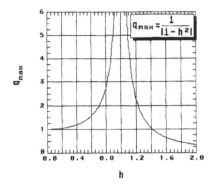

Figure 5.26. *Dynamic amplification factor for $\xi = 0$*

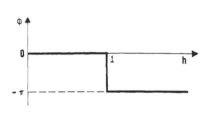

Figure 5.27. *Phase for $\xi = 0$*

The case where $\xi = 0$ is an ideal case: in practice, frictions are never negligible in the vicinity of resonance (apart from resonance, they are sometimes neglected at first approximation to simplify the calculations).

As h varies, q_{max} changes sign while passing through infinity. To preserve at the reduced amplitude the character of an always positive amplitude (the temporal response being symmetrical with respect to the time axis), an abrupt phase shift of value π is introduced into the passage of h = 1.

The phase φ is zero in the interval $0 \leq h \leq 1$; it is then equal to $\pm \pi$ for $h > 1$ (the choice of the sign is uninportatnt). If the value $-\pi$ is taken in $(1, \infty)$, then for example, for $0 \leq h \leq 1$:

$$q_{max} = \frac{\sin h\,\theta}{1 - h^2} \qquad [5.137]$$

and, for $h > 1$:

$$q_{max} = \frac{\sin(h\,\theta - \pi)}{1 - h^2} \qquad [5.138]$$

Particular case where $\xi = 1$

Here

$$q(\theta) = \frac{h}{\left(1 + h^2\right)^2}\left[\frac{1 - h^2}{h}\sin h\,\theta - 2\cos h\,\theta\right] \qquad [5.139]$$

or

$$q(\theta) = H_{RD}(h)\sin\left(h\,\theta - \varphi\right) \qquad [5.140]$$

with

$$H_{RD}(h) = \frac{1}{1 + h^2} \qquad [5.141]$$

and

$$\tan\varphi = \frac{2h}{1 - h^2} \qquad [5.142]$$

NOTE: *The resonance frequency, defined as the frequency for which the response is at a maximum, has the following values:*

Table 5.2. *Resonance frequency and maximum of the transfer function*

Response	Resonance frequency	Amplitude of the relative response
Displacement	$h = \sqrt{1 - 2\xi^2}$	$\dfrac{1}{2\xi\sqrt{1 - \xi^2}}$
Velocity	$h = 1$	$\dfrac{1}{2\xi}$
Acceleration	$h = \dfrac{1}{\sqrt{1 - 2\xi^2}}$	$\dfrac{1}{2\xi\sqrt{1 - \xi^2}}$

(the natural frequency of the system being equal to $h = \sqrt{1 - \xi^2}$ *). For the majority of real physical systems, ξ is small and the difference between these frequencies is negligible.*

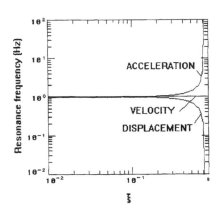

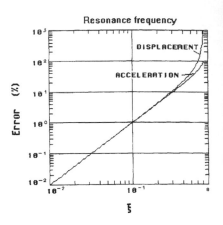

Figure 5.28. *Resonance frequency versus ξ*

Figure 5.29. *Error made by always considering* $h = 1$

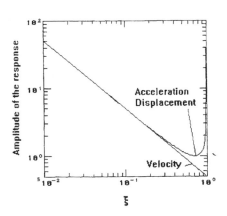

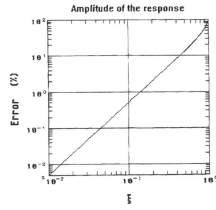

Figure 5.30. *Peak amplitude of the transfer function*

Figure 5.31. *Error made by always taking* $1/2\xi$

5.6. Responses $\dfrac{y}{x_m}, \dfrac{\dot{y}}{\dot{x}_m}, \dfrac{\ddot{y}}{\ddot{x}_m}$ **and** $\dfrac{F_T}{F_m}$

5.6.1. *Movement transmissibility*

Here

$$q(\theta) = H_{AD} \sin(h\,\theta - \varphi) \qquad\qquad [5.143]$$

The maximum amplitude of $q(\theta)$ obtained for $\sin(h\,\theta - \varphi) = 1$, occurring for $h\theta - \varphi = (4\,k + 1)\dfrac{\pi}{2}$, is equal to

$$H_{AD} = \sqrt{\dfrac{1 + 4\,\xi^2\,h^2}{\left(1 - h^2\right)^2 + 4\,\xi^2\,h^2}} \qquad\qquad [5.144]$$

If the excitation is an absolute displacement of the support, the response is the absolute displacement of the mass m. The *movement transmissibility* is defined as the ratio of the amplitude of these two displacements:

$$T_m = \left|\dfrac{y_m}{x_m}\right| \qquad\qquad [5.145]$$

For certain applications, in particular in the case of calculations of vibration isolators or package cushioning, it is more useful to know the fraction of the force amplitude applied to m which is transmitted to the support through the system [BLA 61] [HAB 68]. Then a force transmission coefficient or *force transmissibility* T_f is defined by

$$T_f = \left|\dfrac{F_T}{F_m}\right| \qquad\qquad [5.146]$$

$T_f = T_m = H_{AD}$ is then obtained according to Table 5.1.

5.6.2. *Variations in amplitude*

The amplitude $H_{AD}(h)$ is at a maximum when $\dfrac{dH_{AD}(h)}{dh} = 0$, i.e. for h such that

$$\frac{dH_{AD}}{dh} = \frac{2\,h\left(1 - h^2 - 2\,\xi^2\,h^4\right)}{\sqrt{1 + 4\,\xi^2\,h^2}\left[\left(1 - h^2\right)^2 + 4\,\xi^2\,h^2\right]^{3/2}}$$

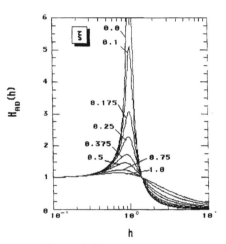

Figure 5.32. *Transmissibility*

This derivative is zero if h = 0 or if

$$1 - h^2 - 2\,\xi^2\,h^4 = 0 \qquad\qquad\qquad [5.147]$$

i.e. for

$$h^2 = \frac{-1 + \sqrt{1 + 8\,\xi^2}}{4\,\xi^2}$$

or, since $h \geq 0$.

$$h = \frac{\sqrt{-1 + \sqrt{1 + 8\,\xi^2}}}{2\,\xi} \qquad\qquad\qquad [5.148]$$

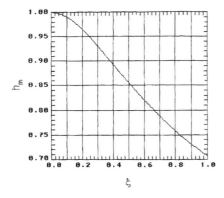

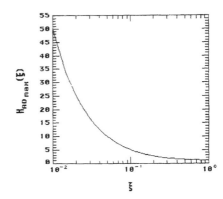

Figure 5.33. *Frequency of the maximum of transmissibility versus* ξ

Figure 5.34. *Maximum of transmissibility versus* ξ

yielding

$$H_{AD_{max}} = \frac{4\,\xi^2}{\sqrt{16\,\xi^4 - 8\,\xi^2 - 2 + 2\,\sqrt{1 + 8\,\xi^2}}} \qquad [5.149]$$

When h tends towards zero, the amplitude H_{AD} tends towards 1 (whatever ξ). When $h \to \infty$, $H_{AD} \to 0$. From the relation [5.147] is drawn

$$\xi^2 = \frac{1 - h^2}{2\,h^4} \qquad [5.150]$$

yielding $h \le 1$.

The locus of the maxima thus has as an equation

$$H_{AD} = \frac{1}{\sqrt{1 - h^4}} \qquad [5.151]$$

This gives the same law as that obtained for relative displacement.

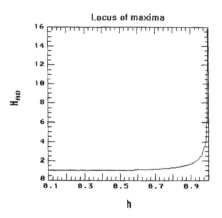

Figure 5.35. *Locus of maximum of transmissibility versus h*

Case of $\xi = 0$

With this assumption, $H_{AD} = H_{RD}$. For all values of ξ, all the curves $\left|H_{AD}(h)\right|$ pass through 1, for $h = 0$ and for $h = \sqrt{2}$. Indeed, $H_{AD}(h) = 1$ if $1 + 4\xi^2 h^2 = \left(1 - h^2\right)^2 + 4\xi^2 h^2$, i.e. $h^2 \left(h^2 - 2\right) = 0$ ($h \geq 0$).

For $h < \sqrt{2}$, all the curves are above $H_{AD} = 1$. Indeed, the condition $1 + 4\xi^2 h^2 > \left(1 - h^2\right)^2 + 4\xi^2 h^2$ is carried out only if $1 > \left(1 - h^2\right)^2$, i.e. if $h < \sqrt{2}$.

In the same way, for $h > \sqrt{2}$, all the curves are below the straight line $H_{AD} = 1$.

5.6.3. *Variations in phase*

If

$$H_{AD}(h) = \left|H_{AD}(h)\right| e^{-j\phi(h)} \qquad\qquad [5.152]$$

$$\tan \phi = \frac{2\xi h^3}{1 - h^2 + 4\xi^2 h^2} \qquad\qquad [5.153]$$

– $\tan \phi = 0$ when $\xi = 0$;

– $\tan \phi \to \infty$ if $\xi = 0$ and $h \to 1$ (thus $\phi \to \dfrac{\pi}{2}$);

– $\tan \phi = 0$ if $h = 0$, i.e. $\phi = 0$;

– $\tan \phi$ behaves like $-\dfrac{2\,\xi\,h}{1 - 4\,\xi^2}$ when $h \to \infty$.

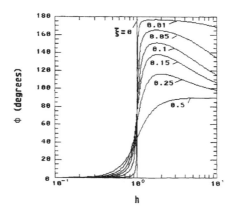

Figure 5.36. *Phase variations*

The denominator is zero if $1 - h^2 + 4\xi^2 h^2 = 0$, i.e. for $h^2 = \dfrac{1}{1 - 4\,\xi^2}$

($\xi < 0.5$) or, since $h \geq 0$,

$$h = \frac{1}{\sqrt{1 - 4\,\xi^2}}$$ [5.154]

In this case, $\tan \phi \to \infty$ and $\phi \to \dfrac{\pi}{2}$.

All the curves have, for $\xi < 1$, a point of inflection at $h = 1$. The slope at this point gets larger as ξ gets smaller.

For $\xi = 0.5$, $\tan \phi = h^3$ ($\phi \to \dfrac{\pi}{2}$ when $h \to \infty$).

For h = 1,

$$\tan \phi = \frac{1}{2\xi}$$ [5.155]

ϕ then gets smaller as ξ gets larger.

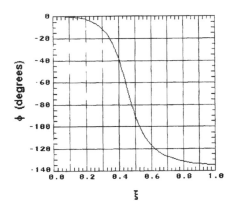

Figure 5.37. *Phase versus ξ for h = 1*

For h $= \dfrac{\sqrt{-1 + \sqrt{1 + 8\,\xi^2}}}{2\,\xi}$.

$$\tan \phi = 2\,\xi \, \frac{\sqrt{1 + 8\,\xi^2} - 1}{\left(2\,\xi - 1\right)\sqrt{1 + 8\,\xi^2} + 1}$$ [5.156]

5.7. Graphical representation of transfer functions

The transfer functions can be plotted in a traditional way on linear or logarithmic axes, but also on a four coordinate nomographic grid, which makes it possible to directly reach the transfer functions of the displacements, the velocities and the accelerations. In this plane diagram at four inputs, the frequency is always carried on the abscissa.

Knowing that $H_{RV} = \Omega\, H_{RD}$ and that $H_{RA} = H_{RV}$, from the ordinate, along the vertical axis the following can be read:

– either the velocity (Figure 5.38). Accelerations are then located on an axis of negative slope (–45°) with respect to the axis of the velocities while the amplitude of the displacements are on an axis at 45° with respect to the same vertical axis. Indeed (Figure 5.39):

$$\log H_{RA} = \log H_{RV} + \log f + \log 2\pi$$

However, a line at 45° with respect to the vertical axis,

$$O'K = O'J + JK = \left(\log H_{RV} + \log 2\pi\right)\frac{\sqrt{2}}{2} + \frac{\sqrt{2}}{2}\log f$$

$$O'K = \frac{\sqrt{2}}{2}\left(\log H_{RV} + \log f + \log 2\pi\right) = \frac{\sqrt{2}}{2}\log H_{RA}$$

$O'K$ is thus proportional to $\log H_{RA}$;

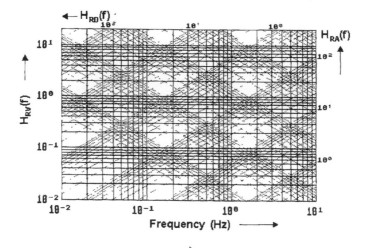

Figure 5.38. *Four coordinate diagram*

– the amplitude of the displacements. A similar calculation shows that the axis of the velocities forms an angle of + 45° with respect to the horizontal line and that of the accelerations an angle of 90° with respect to the axis of the velocities.

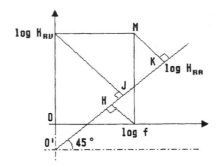

Figure 5.39. *Construction of the four input diagram*

Chapter 6

Non-viscous damping

6.1. Damping observed in real structures

In real structures, damping, which is not perfectly viscous, is actually a combination of several forms of damping. The equation of movement is in consequence more complex, but the definition of damping ratio ξ remains $\dfrac{c}{c_c}$, where c_c is the critical damping of the mode of vibration considered. The exact calculation of ξ is impossible for several reasons [LEV 60]: probably insufficient knowledge of the exact mode of vibration, and of the effective mass of the system, the stiffnesses, the friction of the connections, the constant c and so on. It is therefore important to measure these parameters when possible.

In practice, non-linear damping can often be compared to one of the following categories, which will be considered in the following paragraphs:

– damping force proportional to the power b of the relative velocity $\dot{z}$;

– constant damping force (Coulomb or dry damping), which corresponds to the case where $b = 0$;

– damping force proportional to the square of the velocity ($b = 2$);

– damping force proportional to the square of the relative displacement;

– hysteretic damping, with force proportional to the relative velocity and inversely proportional to the excitation frequency.

Such damping produces a force which is opposed to the direction or the velocity of the movement.

6.2. Linearization of non-linear hysteresis loops – equivalent viscous damping

Generally, the differential equation of the movement can be written [DEN 56]:

$$m \frac{d^2 z}{dt^2} + f(z, \dot{z}) + k\, z = \begin{cases} F_m \sin \Omega\, t \\ -m\, \ddot{x}(t) \end{cases} \qquad [6.1]$$

with, for viscous damping, $f(z, \dot{z}) = c\, \dot{z}$. Because of the presence of this term, the movement is no longer harmonic in the general case and the equation of the movement is no longer linear. such damping leads to nonlinear equations which make calculations complex in a way seldom justified by the result obtained.

Except in some particular cases, such as the Coulomb damping, there is no exact solution. The solution of the differential equation must be carried out numerically. The problem can sometimes be solved by using a development of the Fourier series of the damping force [LEV 60].

Very often in practice damping is fortunately rather weak so that the response can be approached using a sinusoid. This makes it possible to go back to a linear problem, which is easier to treat analytically, by replacing the term $f(z. \dot{z})$ by a force of viscous damping equivalent $c_{eq}\, \dot{z}$; by supposing that the movement response is sinusoidal, the *equivalent damping constant* c_{eq} of a system with viscous damping is calculated which would dissipate the same energy per cycle as nonlinear damping.

The practice therefore consists of determining the nature and the amplitude of the dissipation of energy of the real damping device, rather than substituting in the mathematical models with a viscous damping device having a dissipation of equivalent energy [CRE 65]. It is equivalent of saying that the hysteresis loop is modified.

In distinction from structures with viscous damping, nonlinear structures have non-elliptic hysteresis loops $F_d(z)$ whose form approaches, for example, those shown in Figures 6.1 and 6.2 (dotted curve).

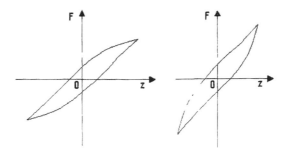

Figure 6.1. *Hysteresis loops of non-linear systems*

Linearization results in the transformation of the real hysteresis loop into an equivalent *ellipse* (Figure 6.2) [CAU 59], [CRE 65], [KAY 77] and [LAZ 68].

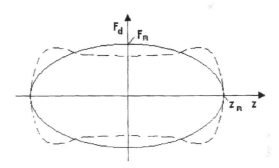

Figure 6.2. *Linearization of hysteresis loop*

Equivalence thus consists in seeking the characteristics of a viscous damping which include:

– the surface delimited by the cycle $F_d(z)$ (same energy dissipation);

– the amplitude of the displacement z_m.

The curve obtained is equivalent only for the selected criteria. For example, the remanent deformation and the coercive force are not exactly the same. Equivalence leads to results which are much better when the non-linearity of the system is lower.

This method, developed in 1930 by L. S. Jacobsen [JAC 30], is general in application and its author was able to show good correlation with the computed results carried out in an exact way when such calculations are possible (*Coulomb damping [DEN 30a]*) and with experimental results. This can, in addition, be extended to the case of systems with several degrees of freedom.

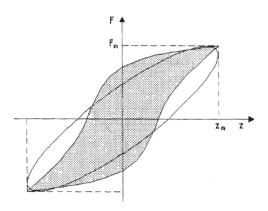

Figure 6.3. *Linearization of hysteresis loop*

If the response can be written in the form $z(t) = z_m \sin(\Omega t - \varphi)$, the energy dissipated by cycle can be calculated using

$$\Delta E_d = \int_{1 \text{ cycle}} F \, dz = \int_{1 \text{ cycle}} f(z, \dot{z}) \frac{dz}{dt} \, dt \qquad [6.2]$$

$$\Delta E_d = z_m \, \Omega \int_0^{2\pi/\Omega} f(z, \dot{z}) \cos(\Omega t - \varphi) \, dt \qquad [6.3]$$

$$\Delta E_d = 4 \, z_m \, \Omega \int_0^{\pi/2\Omega} f(z, \dot{z}) \cos(\Omega t - \varphi) \, dt \qquad [6.4]$$

Energy ΔE_d is equal to that dissipated by an equivalent viscous damping c_{eq} if [HAB 68]:

$$\Delta E_d = c_{eq} \, \Omega \, \pi \, z_m^2 = 4 \, z_m \, \Omega \int_0^{\pi/2\Omega} f(z, \dot{z}) \cos(\Omega t - \varphi) \, dt \qquad [6.5]$$

i.e. if [BYE 67], [DEN 56], [LAZ 68] and [THO 65a]:

$$c_{eq} = \frac{4}{\pi \, z_m} \int_0^{\pi/2\Omega} f(z, \dot{z}) \cos(\Omega t - \varphi) \, dt = \frac{\Delta E_d}{\Omega \, \pi \, z_m^2} \qquad [6.6]$$

The transfer function of a one-degree-of-freedom system $\dfrac{\omega_0^2 \, z_m}{\ddot{x}_m}$ (or in a more

general way $\dfrac{z_m}{F_m / k} = \dfrac{z_m}{\ell_m}$) can be written while replacing c_{eq} by this value in the

relation established, for viscous damping:

$$\frac{z_m}{\ell_m} = \frac{1}{\sqrt{\left(1 - h^2\right)^2 + \left(\dfrac{c_{eq} \, \Omega}{k}\right)^2}} \qquad [6.7]$$

(since $\dfrac{4 \, \xi^2}{\omega_0^2} = \dfrac{c^2}{k^2}$) and for the phase

$$\tan \varphi = \frac{c_{eq} \, \Omega}{k \left(1 - h^2\right)} \qquad [6.8]$$

($h = \dfrac{\Omega}{\omega_0}$).

In addition, $c_{eq} \dfrac{\Omega}{k} = \dfrac{c_{eq}}{k} \dfrac{\Omega}{\omega_0} \omega_0 = \dfrac{c_{eq}}{\sqrt{k \, m}} \, h = 2 \, \xi_{eq} \, h$,

yielding

$$\xi_{eq} = \frac{c_{eq} \, \Omega}{2 \, k \, h} = \frac{c_{eq} \, \omega_0}{2 \, k} \qquad [6.9]$$

$$\xi_{eq} = \frac{\Delta E_d}{2 \, \pi \, h \, k \, z_m^2} \qquad [6.10]$$

$$\frac{z_m}{\ell_m} = \frac{1}{\sqrt{\left(1 - h^2\right)^2 + \left(2 \, \xi_{eq} \, h\right)^2}} \qquad [6.11]$$

If ΔE_d is the energy dissipated by the cycle, the amplitude of the equivalent force applied is [CLO 80]·

$$F_m = \frac{\Delta E_d}{\pi z_m} \tag{6.12}$$

6.3. Main types of damping

6.3.1. *Damping force proportional to the power* b *of the relative velocity*

Table 6.1. *Expressions for damping proportional to power* b *of relative velocity*

Damping force	$F_d = \beta \lvert \dot{z} \rvert^b \dfrac{\dot{z}}{\lvert \dot{z} \rvert}$ or $F_d = \beta \lvert \dot{z} \rvert^b \operatorname{sgn}(\dot{z})$
Equation of the hysteresis loop	$\begin{cases} z = z_m \sin(\Omega t - \varphi) \\ F_d = \beta \left[\Omega z_m \cos(\Omega t - \varphi) \right]^b \operatorname{sgn}(\dot{z}) \end{cases}$ $\dfrac{F_d}{\beta \Omega^b z_m^b} = \operatorname{sgn}(\dot{z}) \left(1 - \dfrac{z^2}{z_m^2} \right)^{b/2}$
Energy dissipated by damping during a cycle	$\Delta E_d = \pi \beta \gamma_b \Omega^b z_m^{b+1}$ with $\gamma_b = \dfrac{2}{\sqrt{\pi}} \dfrac{\Gamma\left(1 + \dfrac{b}{2}\right)}{\Gamma\left(1 + \dfrac{b+1}{2}\right)}$
Equivalent viscous damping	$c_{eq} = \beta \gamma_b \Omega^{b-1} z_m^{b-1}$
Equivalent damping ratio	$\xi_{eq} = \dfrac{\beta z_m^{b-1} \gamma_b h^{b-1} \omega_0^b}{2 k}$
Amplitude of the response	Obeys $z_m^{2b} + \dfrac{\left(1 - h^2\right)^2 \ell_m^{2(b-1)}}{\rho_b^2 h^{2b}} z_m^2 - \dfrac{\ell_m^{2b}}{\rho_b^2 h^{2b}} = 0$ where $\rho_b = \beta \gamma_b \omega_0^b k^{-1} \ell_m^{b-1}$

Phase of the response	$\tan \varphi = \dfrac{\rho_b \; h^b \; z_m^{b-1}}{\ell_m^{b-1} \left(1 - h^2\right)}$

References in Table 6.1: [DEN 30b], [GAM 92], [HAB 68], [JAC 30], [JAC 58], [MOR 63a], [PLU 59], [VAN 57] and [VAN 58].

Relation between b and the parameter J the B. J. Lazan expression

It has been shown [JAC 30] [LAL 96] that if the stress is proportional to the relative displacement z_m ($\sigma = K \; z_m$), the coefficient J of the relation of B.J. Lazan ($D = J \; \sigma^n$) is related to the parameter b by

$$J = \frac{\pi \; \gamma_b \; \beta \; \omega_0^b}{K} \qquad\qquad [6.13]$$

J depends on parameters related to the dynamic behaviour of the structure being considered (K and ω_0).

6.3.2. *Constant damping force*

If the damping force opposed to the movement is independent of displacement and velocity, the damping is known as *Coulomb* or *dry damping*. This damping is observed during friction between two surfaces (dry friction) applied one against the other with a normal force N (mechanical assemblies). It is [BAN 77], [BYE 67], [NEL 80] and [VOL 65]:

– a function of the materials in contact and of their surface quality:

– proportional to the force normal to the interface;

– mainly independent of the relative velocity of slipping between two surfaces;

– larger before the beginning of the relative movement than during the movement in steady state mode.

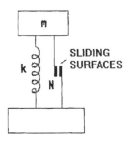

Figure 6.4. *One-degree-of-freedom system with dry friction*

The difference between the coefficients of static and dynamic friction is in general neglected and force N is supposed to be constant and independent of the frequency and of the displacement.

A one-degree-of-freedom system damped by dry friction is represented in Figure 6.4.

Table 6.2. *Expressions for a constant damping force*

Damping force	$F_d = \mu\, N\, \text{sgn}(\dot{z})$
Equation of the hysteresis loop	$F_d = \pm\mu\, N \quad (|z| \le z_m)$
Energy dissipated by damping during a cycle	$\Delta E_d = 4\, z_m\, \mu\, N$
Equivalent viscous damping	$c_{eq} = \dfrac{4\,\mu\, N}{\pi\, z_m\, \Omega}$
Equivalent damping ratio	$\xi_{eq} = \dfrac{2}{\pi}\dfrac{\mu\, N}{k\, h\, z_m}$
Amplitude of the response	$H = \dfrac{z_m}{\ell_m} = \dfrac{1}{\left\|1 - h^2\right\|}\sqrt{1 - \rho_0^2} \qquad \rho_0 = \dfrac{4}{\pi}\dfrac{\mu\, N}{k\, \ell_m} = \begin{cases} \dfrac{4\,\mu\, N}{\pi\, F_m} \\[2ex] -\dfrac{4\,\mu\, N}{\pi\, k\, \ddot{x}_m} \end{cases}$

Phase of the response	$\tan \varphi = \dfrac{\rho_0}{\sqrt{1 - \rho_0^2}}$

[BEA 80], [CRE 61], [CRE 65], [DEN 29], [DEN 56], [EAR 72], [HAB 68], [JAC 30], [JAC 58], [LEV 60], [MOR 63b], [PAI 59], [PLU 59], [ROO 82], [RUZ 57], [RUZ 71], [UNG 73] and [VAN 58].

6.3.3. *Damping force proportional to the square of velocity*

A damping of this type is observed in the case of a body moving in a fluid (applications in fluid dynamics, the force of damping being of the form $C_x \, \rho \, A \, \dfrac{\dot{z}^2}{2}$) or during the turbulent flow of a fluid through an orifice (with high velocities of the fluid, from 2 to 200 m/s, resistance to the movement ceases to be linear with the velocity). When the movement becomes fast [BAN 77], the flow becomes turbulent and the resistance non-linear. Resistance varies with the square of the velocity [BAN 77], [BYE 67] and [VOL 65].

Table 6.3. *Expressions for quadratic damping*

Damping force	$F_d = \beta \, \dot{z} \, |\dot{z}|$ or $F_d = \beta \, \dot{z}^2 \, \mathrm{sgn}(\dot{z})$
Equation of the hysteresis loop	$\dfrac{z^2}{z_m^2} + \dfrac{F_d}{F_{dm}} = 1$
Energy dissipated by damping during a cycle	$\Delta E_d = \dfrac{8}{3} \beta \, \Omega^2 \, z_m^3$
Equivalent viscous damping	$c_{eq} = \dfrac{8 \, \beta \, \Omega \, z_m}{3 \, \pi}$
Equivalent damping ratio	$\xi_{eq} = \rho_2 \, \dfrac{h \, z_m}{2 \, \ell_m}$

Amplitude of the response	$$z_m = \pm \frac{\ell_m \sqrt{-\left(1 - h^2\right)^2 + \sqrt{\left(1 - h^2\right)^4 + 4\,\rho_2^2\,h^4}}}{\sqrt{2}\,\rho_2\,h^2}$$ $$\rho_2 = \beta\,\frac{8}{3\,\pi}\,\frac{\omega_0^2}{k}\,\ell_m$$
Phase of response	$$\tan\,\varphi = \frac{\rho_2\,h^2}{1 - h^2}\sqrt{\frac{2}{\sqrt{\left(1 - h^2\right)^4 + 4\,\rho_2^2\,h^4} + \left(1 - h^2\right)^2}}$$

[CRE 65], [HAB 68], [JAC 30], [RUZ 71], [SNO 68] and [UNG 73].

The constant β is termed the *quadratic damping coefficient*. It is characteristic of the geometry of the damping device and of the properties of the fluid [VOL 65].

6.3.4. *Damping force proportional to the square of displacement*

Table 6.4. *Expressions for damping force proportional to the square of the displacement*

Damping force	$F_d = \gamma\,z^2\,\dfrac{z}{|z|}$ or $F_d = \gamma\,z^2\,\text{sgn}(\dot{z})$
Equation of hysteresis loop	$\begin{cases} z(t) = z_m\,\sin(\Omega\,t - \varphi) \\ F_d(t) = \gamma\,z^2\,\text{sgn}(\dot{z}) = \gamma\,z_m^2\,\sin^2(\Omega\,t - \varphi)\,\text{sgn}(\dot{z}) \end{cases}$
Energy dissipated by damping during a cycle	$\Delta E_d = \pi\,\Omega\,c_{eq}\,z_m^2 = \dfrac{4}{3}\,\gamma\,z_m^3$
Equivalent viscous damping	$c_{eq} = \dfrac{4\,\gamma\,z_m}{3\,\pi\,\Omega}$
Equivalent damping ratio	$\xi_{eq} = \dfrac{2\,\gamma\,z_m}{3\,\pi\,k\,h}$ $\xi_{eq} = \dfrac{4\,\gamma}{3\,\pi}\,\dfrac{\ell_m}{k}\,\dfrac{z_m}{2\,\ell_m\,h} = \dfrac{\theta\,z_m}{2\,\ell_m\,h}$

Amplitude of response	$$z_m^2 = \frac{-\left(1-h^2\right)^2\left(\frac{3\,\pi}{4}\,\frac{k}{\gamma}\right)^2 + \sqrt{\left(\frac{3\,\pi}{4}\,\frac{k}{\gamma}\right)^4\left(1-h^2\right)^4 + 4\,\ell_m^2\left(\frac{3\,\pi}{4}\,\frac{k}{\gamma}\right)^2}}{2}$$
Phase of response	$$\tan\varphi = \frac{\theta}{1-h^2}\ \frac{\sqrt{2}}{\sqrt{\left(1-h^2\right)^2 + \sqrt{\left(1-h^2\right)^4 + 4\,\theta^2}}}$$ $$\beta = \frac{4\,\gamma}{3\,\pi\,k} \qquad \theta = \beta\,\ell_m$$

Such damping is representative of the internal damping of materials, of the structural connections, and cases where the specific energy of damping can be expressed as a function of the level of stress, independent of the form and distribution of the stresses and volume of the material [BAN 77], [BYE 67], [KIM 26] and [KIM 27].

6.3.5. Structural or hysteretic damping

Table 6.5. *Expressions for structural damping*

	Damping coefficient function of Ω	Damping force proportional to the displacement	Complex stiffness
Damping force	$F_d = \dfrac{a}{\Omega}\,\dot{z}$	$F_d = d\,\left\|\dfrac{z}{\dot{z}}\right\|\dot{z} = d\,\|z\|\,\mathrm{sgn}(\dot{z})$	$F = k^*z = \left(k + i\,a\right)z$ or $F = k^*\,z = k\left(1 + i\,\eta\right)z$
Equation of the hysteresis loop	$\dfrac{z^2}{z_m^2} + \dfrac{F_d^2}{a^2 z_m^2} = 1$	$\dfrac{z^2}{z_m^2} + \dfrac{\pi^2 F_d^2}{4\,d^2\,z_m^2} = 1$	$\left\|F^*\right\| = k\,z \pm a\,\sqrt{z_m^2 - z^2}$
Energy dissipated by damping during a cycle	$\Delta E_d = \pi\,a\,z_m^2$	$\Delta E_d = 2\,d\,z_m^2$	$\Delta E_d = \pi\,k\,\eta\,z_m^2$ $\left(= \pi\,a\,z_m^2\right)$
Equivalent viscous damping	$c_{eq} = \dfrac{a}{\Omega}$	$c_{eq} = \dfrac{2\,d}{\pi\,\Omega}$	$c_{eq} = \dfrac{k\,\eta}{\Omega}\ \left(= \dfrac{a}{\Omega}\right)$

Equivalent damping ratio	$\xi_{eq} = \dfrac{a}{2\,m\,\omega_0^2}$	$\xi_{eq} = \dfrac{d}{\pi\,m\,\omega_0^2}$	$\xi_{eq} = \dfrac{a}{2\,k\,h} = \dfrac{\eta}{2\,h}$
Amplitude of response	$z_m = \dfrac{F_m}{k\sqrt{\left(1-h^2\right)^2 + \dfrac{a^2}{k^2}}}$		
Phase of response	$\varphi = \arctan\dfrac{2a/k}{1-h^2}$		

This kind of damping is observed when the elastic material is imperfect when in a system the dissipation of energy is mainly obtained by deformation of material and slip, or friction in the connections. Under a cyclic load, the curve σ, ε of the material forms a closed hysteresis loop rather than only one line [BAN 77]. The dissipation of energy per cycle is proportional to the surface enclosed by the hysteresis loop. This type of mechanism is observable when repeated stresses are applied to an elastic body causing a rise in temperature of the material.

This is called internal friction, hysteretic damping, structural damping or displacement damping. Various formulations are used [BER 76], [BER 73], [BIR 77], [BIS 55], [CLO 80], [GAN 85], [GUR 59], [HAY 72], [HOB 76], [JEN 59], [KIM 27], [LAL 75], [LAL 80], [LAZ 50], [LAZ 53], [LAZ 68], [MEI 67], [MOR 63a], [MYK 52], [PLU 59], [REI 56], [RUZ 71], [SCA 63], [SOR 49] and [WEG 35].

6.3.6 Combination of several types of damping

If several types of damping, as is often the case, are simultaneously present combined with a linear stiffness [BEN 62] [DEN 30a], equivalent viscous damping can be obtained by calculating the energy ΔE_{d_i} dissipated by each damping device and by computing c_{eq} [JAC 30] [JAC 58]:

$$c_{eq} = \frac{\sum_i \Delta E_{d_i}}{\pi\,\Omega\,z_m^2} \qquad\qquad [6.14]$$

Example

Viscous damping and Coulomb damping [JAC 30], [JAC 58], [LEV 60] and [RUZ 71]

$$z = z_m \sin\left(\Omega t - \varphi\right)$$

$$z_m = \frac{\left\{ F_m^2 \left[c^2 \Omega^2 + \left(k - m \Omega^2\right)^2 \right] - \frac{16}{\pi^2} F^2 \left(k - m \Omega^2\right)^2 \right\}^{1/2} - \frac{4}{\pi} c F \Omega}{c^2 \Omega^2 + \left(k - m \Omega^2\right)^2}$$

[6.15]

$$\tan \varphi = \frac{\frac{4}{\pi} F z_m^{-1} \Omega^{-1} + c}{k - m \Omega^2} \Omega$$

[6.16]

F_m = maximum $F(t)$ (excitation)
F = frictional force
c = viscous damping ratio
Ω = pulsation of the excitation

$$\boxed{c_{eq} = \frac{4}{\pi} F z_m^{-1} \Omega^{-1} + c}$$

[6.17]

6.3.7. *Validity of simplification by equivalent viscous damping*

The cases considered above do not cover all the possibilities, but are representative of many situations.

The viscous approach supposes that although non-linear mechanisms of damping are present their effect is relatively small. It is thus applicable if the term for viscous damping is selected to dissipate the same energy by cycle as the system with non-linear damping [BAN 77]. Equivalent viscous damping tends to underestimate the energy dissipated in the cycle and the amplitude of a steady state forced vibration: the real response can larger than envisaged with this simplification.

The decrease of the transient vibration calculated for equivalent viscous damping takes a form different from that observed with Coulomb damping, with a damping force proportional to the square of the displacement or with structural damping. This difference should not be neglected if the duration of the decrease of the response is an important parameter in the problem being considered.

The damped natural frequency is itself different in the case of equivalent viscous damping and in the non-linear case. But this difference is in general so small that it can be neglected.

When damping is sufficiently small (10%), the method of equivalent viscous damping is a precise technique for the approximate solution of non-linear damping problems.

6.4. Measurement of damping of a system

All moving mechanical systems dissipate energy. This dissipation is often undesirable (in an engine, for example), but can be required in certain cases (vehicle suspension, isolation of a material to the shocks and vibrations and so on).

Generally, mass and stiffness parameters can be calculated quite easily. It is much more difficult to evaluate damping by calculation because of ignorance of the concerned phenomena and difficulties in their modelling. It is thus desirable to define this parameter experimentally.

The methods of measurement of damping in general require the object under test to be subjected to vibration and to measure dissipated vibratory energy or a parameter directly related to this energy. Damping is generally studied through the properties of the response of a one-degree-of-freedom mass–spring–damping system [BIR 77] [CLO 80] [PLU 59]. There are several possible methods for evaluating the damping of a system:

– amplitude of the response or amplification factor;

– quality factor;

– logarithmic decrement;

– equivalent viscous damping;

– complex module;

– bandwidth $\dfrac{\Delta f}{f}$.

6.4.1. *Measurement of amplification factor at resonance*

The damping of the one-degree-of-freedom system tends to reduce the amplitude of the response to a sine wave excitation. If the system were subjected to no external forces, the oscillations response created by a short excitation would attenuate and disappear in some cycles. So that the response preserves a constant amplitude, the excitation must bring a quantity of energy equal to the energy dissipated by damping in the system.

The amplitude of the velocity response $\dot{z}$ is at a maximum when the frequency of the sinewave excitation is equal to the resonance frequency f_0 of the system. Since the response depends on the damping of the system, this damping can be deduced from measurement of the amplitude of the response, the one degree of linear freedom system supposedly being linear:

$$Q = \frac{\omega_0 \, \dot{z}_m}{\ddot{x}_m} \tag{6.18}$$

or

$$Q = \frac{\sqrt{k \, m} \, \dot{z}}{F_m} \tag{6.19}$$

For sufficiently small ξ, it has been seen that with a weak error, the amplification factor, defined by $H_{RD} = \dfrac{\omega_0^2 \, z_m}{\ddot{x}_m}$, was equal to Q. The experimental determination of ξ can thus consist of plotting the curve H_{RD} or H_{RV} and of calculating ξ from the peak value of this function. If the amplitude of the excitation is constant, the sum of potential and kinetic energies is constant. The stored energy is thus equal to the maximum of one or the other; it will be, for example $U_s = \dfrac{1}{2} k \, z_m^2$. The energy dissipated during a cycle is equal to [5.87] $\Delta E_d = \pi c \Omega z_m^2$, yielding, since it is supposed that $\Omega = \omega_0$:

$$\frac{U_s}{\Delta E_d} = \frac{1}{2} \frac{k \, z_m^2}{\pi c \omega_0 \, z_m^2} = \frac{k}{2 \pi c \omega_0} = \frac{k \, Q \sqrt{m}}{2 \pi \sqrt{k \, m} \sqrt{k}} \tag{6.20}$$

$$\frac{U_s}{\Delta E_d} = \frac{Q}{2 \pi} \tag{6.21}$$

i.e.

$$Q = \frac{2 \pi U_s}{\Delta E_d} \tag{6.22}$$

NOTE: *The measurement of the response/excitation ratio depends on the configuration of the structure as much as the material. The system is therefore characterized by this rather than the basic properties of the material. This method is not applicable to non-linear systems, since the result is a function of the level of excitation.*

6.4.2. Bandwidth or $\sqrt{2}$ method

Another evaluation method (known as Kennedy–Pancu [KEN 47]) consists of measuring the bandwidth Δf between the half-power points relating to one peak of the transfer function [AER 62], with the height equal to the maximum of the curve H_{RD} (or H_{RV}) divided by $\sqrt{2}$ (Figure 6.5).

From the curve $H_{RV}(h)$, we will have if h_1 and h_2 are the abscissae of the half-power points:

$$Q = \frac{1}{2\,\xi} = \frac{f_0}{\Delta f} \qquad\qquad [6.23]$$

where (f_0 = peak frequency, $h_1 = \dfrac{f_1}{f_0}$, $h_2 = \dfrac{f_2}{f_0}$)

and

$$\xi = \frac{c}{c_c} = \frac{\Delta f}{2\,f_0} \quad \left(= \frac{1}{2}\left(h_2 - h_1\right)\right) \qquad\qquad [6.24]$$

If T_0 is the natural period and T_1 and T_2 are the periods corresponding to an attenuation of $\dfrac{\sqrt{2}}{2}$, damping c is given by

$$c = 2\,\pi\,m \left(\frac{1}{T_2} - \frac{1}{T_1} \right) \qquad\qquad [6.25]$$

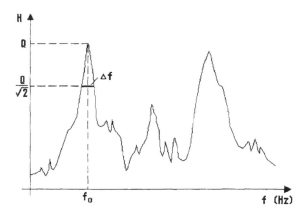

Figure 6.5. *Bandwidth associated with resonance*

since $c_c = 2\sqrt{km}$ and $k = m\,\omega_0^2$, and

$$\boxed{\xi = \frac{T_0\left(T_2 - T_1\right)}{2\,T_1\,T_2}}$$ [6.26]

i.e., with the approximation $f_0 \approx \dfrac{f_1 + f_2}{2}$,

$$\xi \approx \frac{f_2 - f_1}{f_1 + f_2}$$ [6.27]

From the curve H_{RD}, these relations are valid only if ξ is small. The curve H_{AD} could also be used for small ξ.

6.4.3. *Decreased rate method (logarithmic decrement)*

The precision of the method of the bandwidth is often limited by the non-linear behaviour of the material or the reading of the curves. Sometimes it is better to use the traditional relation of logarithmic decrement, defined from the free response of the system after cessation of the exciting force (Figure 6.6).

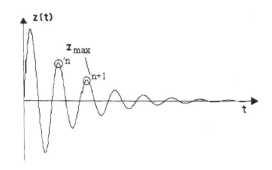

Figure 6.6. *Measurement of logarithmic decrement [BUR 59]*

The amplitude ratio of two successive peaks allows the calculation of the logarithmic decrement δ from

$$\frac{\left(z_m\right)_{n+1}}{\left(z_m\right)_n} = e^{-\delta} \qquad [6.28]$$

In addition, the existence of the following relation between this decrement and damping ratio is also shown

$$\delta = \frac{2\pi\xi}{\sqrt{1-\xi^2}} \qquad [6.29]$$

The measurement of the response of a one-degree-of-freedom system to an impulse load thus makes it possible to calculate δ or ξ from the peaks of the curve [FÖR 37] and [MAC 58]:

$$\xi = \frac{\delta}{\sqrt{\delta^2 + 4\pi^2}} \qquad [6.30]$$

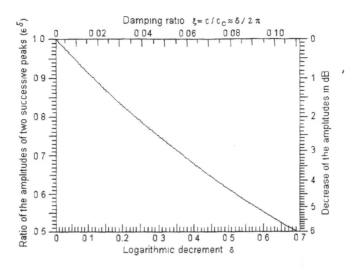

Figure 6.7. *Calculation of damping from* δ

The curve of Figure 6.7 can be use to determine ξ from δ. In order to improve the precision of the estimate of δ, it is preferable to consider two non-consecutive peaks. The definition then used is:

$$\delta = \frac{1}{n-1} \ln \frac{z_{m_1}}{z_{m_n}}$$
[6.31]

where z_{m_1} and z_{m_n} are, respectively, the first and the nth peak of the response (Figure 6.8). In the particular case where ξ is much lower than 1. from [6.29] is obtained:

$$\xi \approx \frac{\delta}{2\pi}$$

yielding

$$\frac{\pi}{\delta} \approx Q$$

and

$$\frac{\pi}{\delta} \approx \frac{2\pi U_{ts}}{D}$$
[6.32]

with

$$\delta = \ln \frac{z_{m1}}{z_{m2}} = \frac{1}{n} \ln \frac{z_{m1}}{z_{mn+1}}$$ [6.33]

with ξ being small

$$\frac{z_{m1}}{z_{mn+1}} \approx 1 + n\,\delta = 1 + 2\,\pi\,n\,\xi_a$$ [6.34]

yielding the approximate value ξ_a

$$\xi_a \approx \frac{z_1 - z_{mn+1}}{2\,\pi\,n\,z_{mn+1}}$$ [6.35]

The error caused by using this approximate relation can be evaluated by plotting the curve $\dfrac{\xi_a - \xi}{\xi}$ according to ξ (Figure 6.8) or that giving the exact value of ξ according to the approximate value ξ_a (Figure 6.9). This gives

$$\xi_a = \frac{z_{m1} - z_{m2}}{2\,\pi\,z_{mm2}} = \frac{1}{2\,\pi} \left[\frac{z_{m1}}{z_{m2}} - 1 \right]$$ [6.36]

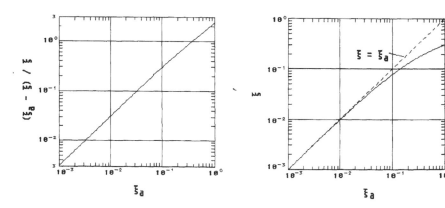

Figure 6.8. *Error related to the approximate relation $\xi(\delta)$*

Figure 6.9. *Exact value of ξ versus approximate value ξ_a*

$$\xi = \frac{\delta}{\sqrt{\delta^2 + 4\pi^2}} = \frac{\ln \dfrac{z_{m1}}{z_{m2}}}{\sqrt{\ln^2 \dfrac{z_{m1}}{z_{m2}} + 4\pi^2}} \qquad [6.37]$$

yielding

$$\xi = \frac{\ln\left(1 + 2\pi\,\xi_a\right)}{\sqrt{\ln^2\left(1 + 2\pi\,\xi_a\right) + 4\pi^2}} \qquad [6.38]$$

and

$$\frac{\xi - \xi_a}{\xi} = 1 - \frac{\xi_a}{\xi} = 1 - \frac{1}{2\pi\xi}\left[e^{2\pi\xi/\sqrt{1-\xi^2}} - 1\right] \qquad [6.39]$$

The specific damping capacity p, the ratio of the specific energy dissipated by damping to the elastic deformation energy per unit of volume, is thus equal to

$$p(\%) = 100\,\frac{D}{U_{ts}} \approx 200\,\delta \qquad [6.40]$$

In a more precise way, p can be also written

$$p = 100\,\frac{z_{m1}^2 - z_{m\,n+1}^2}{n\,z_{m1}^2} \qquad [6.41]$$

while assuming that U_{ts} is proportional to the square of the amplitude of the response. For a cylindrical test-bar,

$$W = S\,\ell\,U_{ts} = \frac{1}{2}k\,z^2 + \frac{1}{2}m\,\dot{z}^2 \qquad [6.42]$$

(potential energy + kinetic energy)

$$\dot{W} = \frac{1}{2}m\left(z^2 + \Omega^2\,z^2\right) \qquad [6.43]$$

i.e., since $z = z_m \sin\left(\Omega t - \varphi\right)$,

$$W = \frac{1}{2}m\Omega^2\,z_m^2 = \text{constant } z_m^2 \qquad [6.44]$$

U_{ts} is thus proportional to z_m^2 yielding, from [6.31] and [6.41] for two successive peaks:

$$p\ (\%) = 100 \left(1 - e^{-2\delta}\right)$$

[6.45]

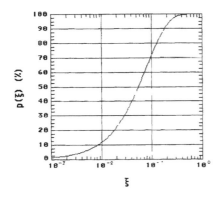

Figure 6.10. *Specific damping capacity versus* ξ

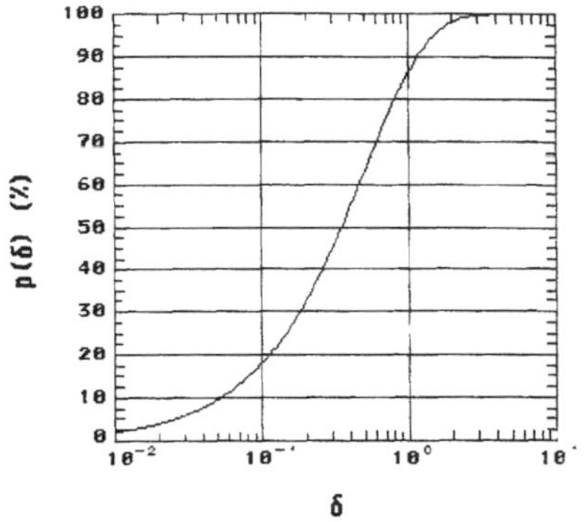

Figure 6.11. *Specific damping capacity versus* δ

The use of the decrement to calculate p from the experimental results supposes that δ is constant during n cycles. This is not always the case. It was seen that damping increases as a power of the stress, i.e. of the deformation and it is thus desirable to use this method only for very low levels of stress.

For δ small, we can write [6.45] in the form of a series:

$$p\ (\%) = 100 \left[\frac{2\,\delta}{1!} - \frac{(2\,\delta)^2}{2!} + \frac{(2\,\delta)^3}{3!} - \cdots \right]$$

[6.46]

If $\delta < 0.01$, we find $p \approx 200\,\delta$.

The method of logarithmic decrement takes no account of the non-linear effects. The logarithmic decrement δ can be also expressed according to the resonance peak amplitude H_{max} and its width Δf at an arbitrary height H [BIR 77], [PLU 59]. F. Förester [FÖR 37] showed that

$$\delta = \pi \frac{\Delta f}{f_0} \sqrt{\frac{H^2}{H_{max}^2 - H^2}}$$

[6.47]

$$\delta = \pi \frac{\Delta f}{f_0} \sqrt{\frac{H^2}{Q^2 - H^2}}$$

[6.48]

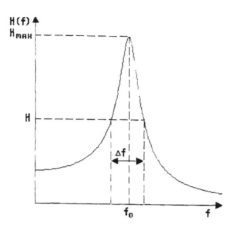

Figure 6.12. *Bandwidth at amplitude* H

If $H = \dfrac{Q}{2}$.

$$\delta = \frac{\pi}{\sqrt{3}} \frac{\Delta f}{f_0}$$

[6.49]

Setting as n_e the number of cycles such that the amplitude decreases by a factor e (number of Neper), it becomes

$$\delta = -\frac{1}{\left(n_e - 1\right)} \ln \frac{1}{e} = \frac{1}{\left(n_e - 1\right)} = \frac{1}{f_0 \, t_e}$$

[6.50]

where t_e = time to reach the amplitude $\dfrac{z_{m1}}{e}$. If the envelope $Z(t)$ of the response $z(t)$ (which is roughly a damped sinusoid) is considered this gives.

$$\delta = \frac{1}{f_0 \, Z} \frac{dZ}{dt} = -\frac{1}{f_0} \frac{d \ln Z}{dt} = -\frac{2.302}{f_0} \frac{d \ln Z}{dt}$$

[6.51]

and if the amplitude in decibels is expressed as

$$y_{dB} = 20 \log Z$$

$$\delta = -\frac{0.115}{f_0} \frac{dy}{dt}$$
[6.52]

For a value of H such that $H^2 = \frac{Q^2}{2}$,

$$\delta = \pi \frac{\Delta f}{f_0}$$
[6.53]

If $\xi \leq 0.1$, $\delta \approx \frac{\pi}{Q}$, yielding $Q = \frac{f_0}{\Delta f}$, a relation already obtained. The calculation of the Q factor from this result and from the curve $H(f)$ can lead to errors if the damping is not viscous.

In addition, it was supposed that the damping was viscous. If this assumption is not checked different values of δ are obtained depending on the peaks chosen, particularly for peaks chosen at the beginning or end of the response [MAC 58].

Another difficulty can arise in the case of a several degrees-of-freedom system for which it can be difficult to excite only one mode. If several modes are excited, the response of a combination of several sinusoids to various frequencies will be presented.

6.4.4. *Evaluation of energy dissipation under permanent sinusoidal vibration*

An alternative method can consist of subjecting the mechanical system to harmonic excitation and to evaluate, during a cycle, the energy dissipated in the damping device [CAP 82], where the quantity is largely accepted as a measure of the damping ratio.

This method can be applied to an oscillator whose spring is not perfectly elastic. this then leads to the constant k and c of an equivalent simple oscillator.

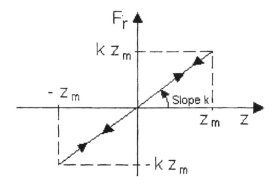

Figure 6.13. *Force in a spring*

It has been seen that, if a one-degree-of-freedom mechanical system is subjected to a sinusoidal force $F(t) = F_m \sin \Omega t$ such that the pulsation is equal to the natural pulsation of the system (ω_0), the displacement response is given by

$$z(t) = -z_m \cos \Omega t$$

where

$$z_m = \frac{F_m}{2 k \xi}$$

The force F_s in the spring is equal to $F_s = k z(t)$ and the force F_d in the damping device to $F_d = c \dot{z} = 2 m \xi \Omega \dot{z} = 2 k \xi z_m \sin \Omega t$, yielding F_d according to z:

$$\frac{F_d^2}{(2 k \xi z_m)^2} = \sin^2 \Omega t = 1 - \frac{z^2}{z_m^2} \qquad [6.54]$$

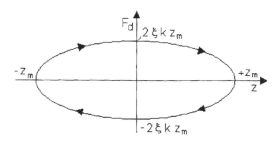

Figure 6.14. *Damping force versus displacement*

This function is represented by an ellipse. During a complete cycle the potential energy stored in the spring is entirely restored. On the other hand, there is energy ΔE_d spent in the damping device, that is equal to the surface of the ellipse:

$$\Delta E_d = 2 \pi z_m^2 k \xi \qquad [6.55]$$

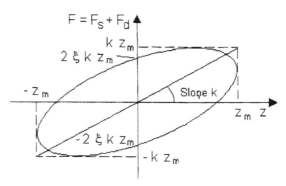

Figure 6.15. *Total force versus elongation*

The superposition of Figures 6.13 and 6.14 makes it possible to plot $F = F_s + F_d$ against z (Figure 6.15).

From these results, the damping constant c is measured as follows:

– by plotting the curve $F(z)$ after moving the system out of equilibrium (force F applied to the mass);

– by taking the maximum deformation z_m .

It is supposed here that the stiffness k is linear and k is thus calculated from the slope of the straight line plotted midway at the centre of the ellipse.

The surface of the ellipse gives ΔE_d , yielding

$$\xi = \frac{\Delta E_d}{2 \pi z_m^2 k} \qquad [6.56]$$

NOTES:

1. *When z_m increases, the spring has an increasingly non-linear behaviour in general and the value of ξ obtained grows.*

2. *The energy dissipated by the cycle* (ΔE_d) *depends on the form, dimensions and the distribution of the dynamic stresses. It is preferable to consider the specific damping energy* D, *which is a basic characteristic of the material (damping energy per cycle and unit of volume by assuming a uniform distribution of the dynamic stresses in the volume* V *considered)* [PLU 59].

$$\Delta E_d = \int_V D \ dV \qquad\qquad [6.57]$$

where ΔE_d *is in Joules/cycle and* D *is in Joules/cycle/m^3.*

Some examples of different values of ξ are given in Table 6.6 [BLA 61] and [CAP 82].

Rubber-type materials, with weak damping

The dynamic properties of Neoprene show a very weak dependence on the frequency. The damping ratio of Neoprene increases more slowly at high frequencies than the damping ratio of natural rubber [SNO 68].

Table 6.6. *Examples of damping values*

Material	ξ
Welded metal frame	0.04
Bolted metal frame	0.07
Concrete	0.010
Prestressed concrete	0.05
Reinforced concrete	0.07
High-strength steel (springs)	$0.637 \ 10^{-3}$ to $1.27 \ 10^{-3}$
Mild steel	$3.18 \ 10^{-3}$
Wood	$7.96 \ 10^{-3}$ to $31.8 \ 10^{-3}$
Natural rubber for damping devices	$1.59 \ 10^{-3}$ to $12.7 \ 10^{-3}$
Bolted steel	0.008
Welded steel	0.005

Rubber-type materials, with strong damping

The dynamic module of these materials increases very rapidly with the frequency. The damping ratio is large and can vary slightly with the frequency.

6.4.5. *Other methods*

Other methods were developed to evaluate the damping of the structures such as, for example, that using the derivative to the resonance of the phase with respect to the frequency (Kennedy-Pancu improved method) [BEN 71].

6.5. Non-linear stiffness

We considered in paragraph 6.3 the influence of non-linear damping on the response of a one-degree-of-freedom system. The non-linearity was thus brought about by damping. Another possibility relates to the non linearities due to the stiffness. It can happen that the stiffness varies according to the relative displacement response. The restoring force, which has the form $F = -k\, z$, is not linear any more and can follow a law such as, for example, $F = k\, z + r\, z^3$ where k is the preceding constant and where r determines the rate of non-linearity. The stiffness can increase with relative displacement (hardening spring) (Figure 6.16) or decrease (softening spring) (Figure 6.17) [MIN 45].

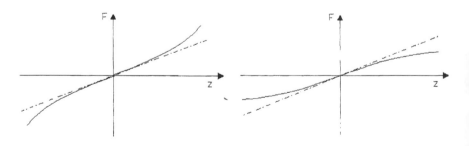

Figure 6.16. *Hardening spring*   **Figure 6.17.** *Softening spring*

A well-known phenomenon of jump [BEN 62] can then be observed on the transfer function.

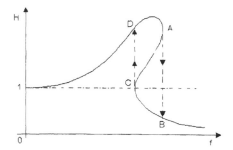

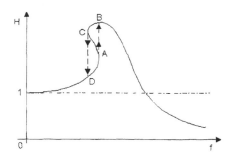

Figure 6.18. *Transmissibility against increasing frequency*

Figure 6.19. *Transmissibility against decreasing frequency*

When the frequency increases slowly from zero, the transmissibility increases from the intercept 1 up to point A while passing through D and then decreases to B (Figure 6.18).

If, on the contrary, resonance is approached from high frequencies by a slow sinusoidal sweeping at decreasing frequency, the transfer function increases, passes through C and moves to D near the resonance, and then decreases up to 1 as f tends towards zero (Figure 6.19).

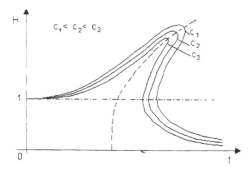

Figure 6.20. *Influence damping*

It should be noted that the area CA is unstable and therefore cannot represent the transfer function of a physical system.

The shape of the curve depends on the amplitude of the force of excitation, like the frequency of resonance. The mass can vibrate at its natural frequency with an excitation frequency much larger (phenomenon known as *resonance of the nth order)* [DUB 59].

Chapter 7

Swept sine

7.1. Definitions

7.1.1. *Swept sine*

A *swept sine* can be defined as a function characterized by a relation of the form:

$$\ell(t) = \ell_m \, \sin[E(t) + \phi] \qquad\qquad [7.1]$$

where

- the phase ϕ is in general zero;

- $E(t)$ is a time function characteristic of the sweep mode;

- $\ell(t)$ is generally an acceleration, sometimes a displacement, a velocity or a force.

The pulsation of the sinusoid can be defined like the derivative of the function under the symbol *sine* [BRO 75], [HAW 64], [HOK 48], [LEW 32], [PIM 62], [TUR 54] and [WHI 72], i.e. by:

$$\Omega = 2\,\pi\,f = \frac{dE}{dt} \qquad\qquad [7.2]$$

We will see that the most interesting sweep modes are:

- the *linear sweep*, for which f of form $f = \alpha\, t + \beta$;

– the *logarithmic sweep* (which should rather be termed exponential) if
$f = f_1 \, e^{t/T_1}$;

– the *hyperbolic sweep* (or parabolic, or log log) if: $\dfrac{1}{f_1} - \dfrac{1}{f} = at$

These sweeps can be carried out at an increasing frequency or a decreasing frequency.

The first two laws are the most frequently used in laboratory tests. Other laws can however be met some of which have been the subject of other published work [SUZ 78a], [SUZ 79], [WHI 72].

7.1.2. Octave - number of octaves in frequency interval (f_1, f_2)

An *octave* is the interval ranging between two frequencies whose ratio is two. The number of octaves ranging between two frequencies f_1 and f_2 is such that:

$$\frac{f_2}{f_1} = 2^n \qquad\qquad [7.3]$$

yielding

$$n = \frac{\ln \dfrac{f_2}{f_1}}{\ln 2} \qquad\qquad [7.4]$$

(logarithms in both cases being base e or base 10).

7.1.3. Decade

A *decade* is the interval ranging between two frequencies whose ratio is ten. The number of decades n_d ranging between two frequencies f_1 and f_2 is such that:

$$\frac{f_2}{f_1} = 10^{n_d} \qquad\qquad [7.5]$$

yielding

$$n_d = \log \frac{f_2}{f_1} = \frac{\ln f_2/f_1}{\ln 10} \qquad\qquad [7.6]$$

($\ln 10 = 2.30258 \ldots$)

The relation between the number of decades and the number of octaves ranging between two frequencies

$$\ln \frac{f_2}{f_1} = n \ln 2 = n_d \ln 10 \qquad\qquad [7.7]$$

$$\frac{n}{n_d} = \frac{\ln 10}{\ln 2} \approx 3.3219\ldots \qquad\qquad [7.8]$$

7.2. 'Swept sine' vibration in the real environment

Such vibrations are relatively rare. They are primarily measured on structures and equipment installed in the vicinity of rotating machines, at times of launching, stop or speed changes. They were more particularly studied to evaluate their effects during transition through the resonance frequency of a material [HAW 64], [HOK 48], [KEV 71], [LEW 32], [SUZ 78a], [SUZ 78b] and [SUZ 79].

7.3. 'Swept sine' vibration in tests

Material tests were and are still very often carried out by applying an excitation of the sine type to the specimen, the objectives being:

– Identification of the material: the test is carried out by subjecting the material to a swept sine having in general a rather low and constant amplitude (not to damage the specimen), about 5 ms^{-2}, the variation of the frequency with time being rather small (close to one octave per minute) in order to study the response at various points of the specimen, to emphasize the resonance frequencies and to measure the amplification factors;

– The application of a test defined in a standard document (MIL STD 810 C, AIR 7304, GAM T 13...), the test being intended to show that the material has a certain standard robustness, independent or difficult to relate to the vibrations which the material will undergo in its service life;

– The application of a specification which as well as being feasible covers vibrations in its future real environment.

The swept sine can have a constant level over all the frequency band studied (Figure 7.1-a) or can be composed of several constant levels at various frequency intervals (Figure 7.1-b).

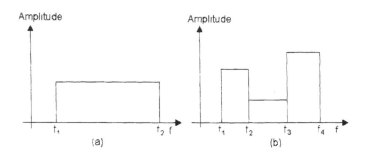

Figure 7.1. *Examples of swept sines*

Sweep is most frequently logarithmic. The specification sometimes specifies the direction of sweep: *increasing* or *decreasing frequency.*

Either *the sweep rate* (number of octaves per minute), or the *sweep duration* (from lowest frequency to highest or in each frequency band) is specified.

The level is defined by the peak value of the sinusoid or the peak-to-peak amplitude. As for the *sine test at constant frequency* (a redundant expression because the definition of a sinusoid includes this assumption; this terminology is however commonly used for better distinguishing these two vibration types. This test is also termed the *dwell test*), the parameter characterizing the amplitude can be a displacement (a displacement is sometimes specified at very low frequencies, this parameter being easier to measure in this frequency domain), more rarely a velocity, and, in general, an acceleration.

The the sweep rate is, in general, selected to be sufficiently low to enable the response of the equipment to the test to reach a very strong percentage of the level obtained in steady operation under pure sinusoidal excitation.

If the sweep is fast, it can be estimated that each resonance is excited one after the other, in a transient way, when the frequency sweeps the interval ranging between the half-power points of each peak of the transfer function of the material. We will see (Volume 3) how this method can be used to measure the transfer functions.

In this direction, the swept sine is a vibration of which the effects can be compared with those of a shock (except the fact that under a shock, all the modes are excited simultaneously) [CUR 55].

The swept sinusoidal vibration tests are adapted badly to simulate random vibrations, whose amplitude and phase vary in a random way and with which all the frequencies are excited simultaneously.

7.4. Origin and properties of main types of sweepings

7.4.1. *The problem*

We know that for a linear one-degree-of-freedom mechanical system the damping ratio is given by:

$$\xi = \frac{c}{2\sqrt{k\,m}}$$
[7.9]

the Q factor by:

$$Q = \frac{1}{2\xi} = \frac{\sqrt{k\,m}}{c} = 2\pi f_0 \frac{m}{c}$$
[7.10]

the resonance frequency by:

$$f_0 = \frac{1}{2\pi}\sqrt{\frac{k}{m}} \left(= \frac{\omega_0}{2\pi} \right)$$
[7.11]

and the width Δf of the peak of the transfer function between the half-power points by:

$$\Delta f = \frac{f_0}{Q}$$
[7.12]

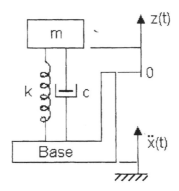

Figure 7.2. *One-degree-of-freedom system*

NOTE: *We saw that the maximum of the transfer function* $\left|H(f)\right| = \dfrac{\omega_0^2 \, z_{max}}{\ddot{x}_{max}}$ *is actually equal to*

$$H_m = \frac{1}{2\,\xi\,\sqrt{1-\xi^2}} = \frac{Q}{\sqrt{1-\xi^2}} \qquad [7.13]$$

The mechanical systems have, in general, rather weak damping so that the approximation $H_m = Q$ *(which is the exact result for the maximum of the transfer function acceleration – relative velocity instead here of the function acceleration - relative displacement) can be used. One additionally recalls that the half-power points are defined because of mechanical-electric analogy, from the transfer function acceleration-relative velocity.*

Writing:

$$\dot{f} = \frac{df}{dt} \qquad [7.14]$$

for the sweep rate around the resonance frequency, f_0, the time spent in the band Δf is given roughly by:

$$\Delta t = \frac{\Delta f}{\dot{f}} \qquad [7.15]$$

and the number of cycles performed by:

$$\Delta N = f_0 \,\Delta t = \frac{\Delta f}{\dot{f}}\, f_0 \qquad [7.16]$$

When such a system is displaced from its equilibrium position and then released (or when the excitation to which it is subjected is suddenly stopped), the displacement response of the mass can be written in the form:

$$z(t) = z_m \, e^{-t/T} \cos\!\left(2\,\pi\,f_0\,\sqrt{1-\xi^2}\;t + \phi\right) \qquad [7.17]$$

where T is a time constant equal to:

$$T = \frac{2\,m}{c} \qquad [7.18]$$

i.e. according to [7.10]:

$$T = \frac{Q}{\pi \, f_0} = \frac{1}{\omega_0 \, \xi} \qquad\qquad [7.19]$$

It will be supposed in the following that the Q factor is independent of the natural frequency f_0, in particular in the case of viscoelastic materials. Different reasoning can take into account a variation of Q with f_0 according to various laws [BRO 75]; this leads to the same laws of sweeping.

In a swept sine test, with the frequency varying according to time the response of a mechanical system is never perfectly permanent. It is closer to the response which the system would have under permanent stress at a given frequency when the sweep rate is slower. To approach as closely as possible this response in the vicinity of the resonance frequency, it is required that the time Δt spent in Δf be long compared to the constant T, a condition which can be written [MOR 76]:

$$\Delta t = \mu \, T \qquad\qquad [7.20]$$

$(\mu \gg 1)$; yielding[1]

$$\left| \ddot{f} \right| = \frac{\Delta f}{\Delta t} = \Delta f \, \frac{\pi \, f_0}{\mu \, Q} = \frac{f_0}{Q} \, \frac{\pi \, f_0}{\mu \, Q} \qquad\qquad [7.21]$$

$$\left| \ddot{f} \right| = \frac{\pi \, f_0}{\mu \, Q^2} \qquad\qquad [7.22]$$

The natural frequency f_0 can be arbitrary in the band considered (f_1, f_2) and, whatever its value, the response must be close to Q times the input to the resonance. To calculate the sweep law $f(t)$ let us generalize f_0 by writing f as:

$$\dot{f} = \pm \frac{\pi \, f^2}{\mu \, Q^2} \qquad\qquad [7.23]$$

It can be seen that the sweep rate varies as $1/Q^2$.

NOTE: *The derivative $\dot{f}$ is positive for increasing frequency sweep, negative for decreasing frequency sweep.*

---

1. It is supposed here that Δf is sufficiently small (i.e. ξ is small) to be able to equalize with little error the slope of the tangent to the curve f(t) and the slope of the chord relating to the interval Δf. We will see that this approximation is indeed acceptable in practice.

7.4.2. Case n°1: sweep where time Δt spent in each interval Δf is constant for all values of f_0

Here, since

$$\Delta t = \mu \, T = \frac{\mu \, Q}{\pi \, f} \qquad\qquad [7.24]$$

it is necessary that $\mu = \gamma \, f$ the constant γ has the dimension of time, and

$$\Delta t = \frac{\gamma \, Q}{\pi} \qquad\qquad [7.25]$$

$$\dot{f} = \pm \frac{\pi \, f^2}{\mu \, Q^2} = \pm \frac{\pi \, f}{\gamma \, Q^2} = \pm \frac{f}{T_1} \qquad\qquad [7.26]$$

if we set $T_1 = \dfrac{\gamma \, Q^2}{\pi}$.

Sweeping at frequency increasing between f_1 and f_2

We deduce from [7.26]

$$f = f_1 \, e^{\frac{t}{T_1}} \qquad\qquad [7.27]$$

The constant T_1 is such that, for $t = t_s$, $f = f_2$:

$$T_1 = \frac{t_s}{\ln f_2 / f_1} \qquad\qquad [7.28]$$

where T_1 is the time needed to sweep the interval between two frequencies whose ratio is e. Relations [7.24] and [7.25] lead to

$$T_1 = Q \, \Delta t \qquad\qquad [7.29]$$

Sweep at decreasing frequency

$$f = f_2 \, e^{-\frac{t}{T_1}}$$

[7.30]

the constant T_1 having the same definition as previously.

Expression for E(t)

Increasing frequency:

$$E(t) = 2\pi \int_0^t f_1 \, e^{t/T_1} \, dt$$

[7.31]

i.e. [HAW 64] and [SUN 75]:

$$E(t) = 2\pi T_1 f_1 \left(e^{t/T_1} - 1 \right) = 2\pi T_1 \left(f - f_1 \right)$$

[7.32]

Decreasing frequency:

$$E(t) = 2\pi \int_0^t f_2 \, e^{-t/T_1} \, dt$$

[7.33]

$$E(t) = -2\pi T_1 f_2 \left(e^{-t/T_1} - 1 \right) = -2\pi T_1 \left(f - f_2 \right)$$

[7.34]

Later in this paragraph, and except in a specified particular case, we will consider only sweepings at increasing frequency, the relations being for the other case either identical or very easy to rewrite.

We supposed above that f_1 is always, whatever the sweep direction, the lowest frequency, and f_2 always the highest frequency. Under this assumption, certain relations depend on the sweep direction. If, on the contrary, it is supposed simply that f_1 is the initial frequency of sweep and f_2 the final frequency, whatever the direction, we obtain the same relations independently of the direction; relations in addition identical to those established above and in what follows in the case of an increasing frequency.

Time t can be expressed versus the frequency f according to:

$$t = T_1 \ln \frac{f}{f_1}$$

[7.35]

In spite of the form of the relations [7.27] and [7.30], the sweep is known as *logarithmic,* by referring to he expression [7.35].

The time necessary to go from frequency f_1 to frequency f_2 is given by:

$$t_s = T_1 \ln \frac{f_2}{f_1} \qquad [7.36]$$

which can still be written:

$$t_s = Q \, \Delta t \, \ln\frac{f_2}{f_1} \qquad [7.37]$$

The number of cycles carried out during time t is given by:

$$N = \int_0^t f(t)\,dt = \int_0^t f_1 \, e^{\frac{t}{T_1}} \, dt \qquad [7.38]$$

$$N = f_1 \, T_1 \left(e^{t/T_1} - 1 \right) \qquad [7.39]$$

i.e., according to [7.27]:

$$N = T_1 \left(f - f_1 \right) \qquad [7.40]$$

The number of cycles between f_1 and f_2 is:

$$N_s = T_1 \left(f_2 - f_1 \right) \qquad [7.41]$$

which can be also written, taking into account [7.36],

$$N_s = \frac{t_s \left(f_2 - f_1 \right)}{\ln\dfrac{f_2}{f_1}} \qquad [7.42]$$

The mean frequency (or *average frequency* or *expected frequency*):

$$f_m = \frac{N_s}{t_s} = \frac{f_2 - f_1}{\ln f_2/f_1} \qquad [7.43]$$

The number of cycles ΔN performed in the band Δf between the half-power points (during time Δt) is written [7.41]:

$$\Delta N = T_1 \left[f_0 \left(1 + \frac{1}{2Q} \right) - f_0 \left(1 - \frac{1}{2Q} \right) \right]$$

i.e.

$$\Delta N = f_0 \frac{T_1}{Q}$$

$$\Delta N = f_0 \ \Delta t \qquad\qquad [7.44]$$

ΔN thus varies like f_0 yielding

$$t_s = \frac{Q \ \Delta N}{f_0} \ \ln \frac{f_2}{f_1} \qquad\qquad [7.45]$$

Starting also from [7.41]:

$$\Delta N = \frac{f_0 \ N_s}{Q (f_2 - f_1)} \qquad\qquad [7.46]$$

The time Δt spent in an interval Δf is:

$$\Delta t = \frac{T_1}{Q}$$

$$\Delta t = \frac{t_s}{Q \ln \dfrac{f_2}{f_1}} \qquad\qquad [7.47]$$

yielding another expression of ΔN :

$$\Delta N = f_0 \ \Delta t$$

$$\Delta N = \frac{f_0 \ t_s}{Q \ln \dfrac{f_2}{f_1}} \qquad\qquad [7.48]$$

The number of cycles N_1 necessary to go from frequency f_1 to a resonance frequency f_0 is:

$$N_1 = T_1 \left(f_0 - f_1 \right) \qquad\qquad [7.49]$$

$$N_1 = Q \ t\left(f_0 - f_1\right) \tag{7.50}$$

$$N_1 = \frac{Q \, \Delta N}{f_0}\left(f_0 - f_1\right) \tag{7.51}$$

or

$$N_1 = \frac{t_s \left(f_0 - f_1\right)}{\ln\dfrac{f_2}{f_1}} \tag{7.52}$$

This number of cycles is carried out in time:

$$t_1 = T_1 \ln \frac{f_0}{f_1} \tag{7.53}$$

$$t_1 = \frac{Q \ \Delta N}{f_0} \ln \frac{f_0}{f_1} \tag{7.54}$$

or

$$t_1 = Q \ \Delta t \ln\frac{f_0}{f_1} = \frac{t_s \ \ln\dfrac{f_0}{f_1}}{\ln\dfrac{f_2}{f_1}} \tag{7.55}$$

If the initial frequency f_1 is zero, we have $N_1 = N_0$ given by:

$$N_0 = f_0 \ T_1 \tag{7.56}$$

or

$$N_0 = Q \ \Delta t \, f_0 = Q \, \Delta N \tag{7.57}$$

It is not possible, in this case, to calculate the time t_0 necessary to go from 0 to f_0.

Sweep rate

According to the sweep direction, we have:

$$\frac{df}{dt} = \begin{Bmatrix} \dfrac{f_1}{T_1} e^{t/T_1} \\[2ex] \dfrac{f_2}{T_1} e^{-t/T_1} \end{Bmatrix} = \frac{f}{T_1} \qquad [7.58]$$

The sweep rate is expressed, generally, in number of octaves per minute. The number of octaves between two frequencies f_1 and f_2 is equal to [7.4]:

$$n = \frac{\ln f_2/f_1}{\ln 2}$$

yielding the number of octaves per second:

$$R_{os} = \frac{n}{t_s} = \frac{\ln f_2/f_1}{t_s \ln 2} \qquad [7.59]$$

(t_s being expressed in seconds) and the number of octaves per minute:

$$R_{om} = \frac{60\,n}{t_s} = 60\,R_{os} \qquad [7.60]$$

$$t_s = \frac{60}{R_{om} \ln 2} \ln \frac{f_2}{f_1}$$

If we set $f = f_2$ in [7.35] for $t = t_s$, we obtain:

$$t_s = T_1 \ln\frac{f_2}{f_1} = \frac{60}{R_{om}} \frac{\ln f_2/f_1}{\ln 2} \qquad [7.61]$$

From [7.59] and from [7.61]:

$$\ln\frac{f_2}{f_1} = \frac{t_s}{T_1} = R_{os}\, t_s \ln 2$$

$$R_{os} = \frac{1}{T_1 \ln 2} \qquad [7.62]$$

yielding another expression of the sweep law:

$$f = f_1 \, 2^{R_{os} t}$$

[7.63]

or

$$f = f_1 \, 2^{R_{om} t / 60}$$

[7.64]

according to the definition of R.

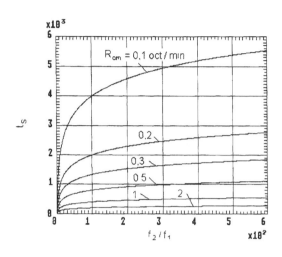

Figure 7.3. *Sweep duration*

Figure 7.3 shows the variations of t_s versus the ratio f_2/f_1, for R_{om} equal to $0.1 - 0.2 - 0.3 - 0.5 - 1$ and 2 Oct.-min (relation [7.60]). In addition we deduce from [7.58] and [7.62] the relation:

$$\frac{df}{dt} = \frac{f}{T_1} = f \, R_{os} \ln 2$$

[7.65]

Case where the sweep rate R_{dm} *is expressed in decades per minute*

By definition and according to [7.6]

$$R_{dm} = \frac{n_d}{t_s / 60} = \frac{60 \ln f_2/f_1}{t_s \ln 10}$$

[7.66]

$$R_{dm} = \frac{60 \ln f_2/f_1}{t_s}$$ [7.67]

or

$$R_{dm} = R_{om} \frac{\ln 2}{\ln 10} \approx \frac{R_{om}}{3.3219...}$$ [7.68]

The time spent between two arbitrary frequencies in the swept interval f_1, f_2

Setting f_A and f_B ($> f_A$) the limits of a frequency interval located in (f_1, f_2). The time $t_B - t_A$ spent between f_A and f_B is calculated directly starting from [7.35] and [7.36] for example:

$$t_B - t_A = t_s \frac{\ln f_B/f_A}{\ln \dfrac{f_2}{f_1}}$$ [7.69]

Time spent between the half-power points of a linear one-degree-of-freedom system

Let us calculate the time Δt^* between the half-power points using the relations established for small ξ.

The half-power points have as abscissae $f_0 - \dfrac{\Delta f}{2}$ and $f_0 + \dfrac{\Delta f}{2}$ respectively, i.e.

$f_0 \left(1 - \dfrac{1}{2Q}\right)$ and $f_0 \left(1 + \dfrac{1}{2Q}\right)$, yielding, starting from [7.69], the time Δt^* spent between these points:

$$\Delta t^* = t_s \frac{\ln \dfrac{1 + 1/2\,Q}{1 - 1/2\,Q}}{\ln f_2/f_1}$$ [7.70]

This relation can be written:

$$\Delta t^* = T_1 \ln \frac{1 + \xi}{1 - \xi}$$

i.e., since $\Delta t = T_1/Q$

$$\frac{\Delta t^*}{\Delta t} = \frac{1}{2\,\xi}\,\ln\frac{1+\xi}{1-\xi} \qquad\qquad [7.71]$$

Figure 7.4 shows the variations of $\dfrac{\Delta t^*}{\Delta t}$ versus ξ. It is noted that, for $\xi < 0.2$,

$\dfrac{\Delta t^*}{\Delta t}$ is very close to 1.

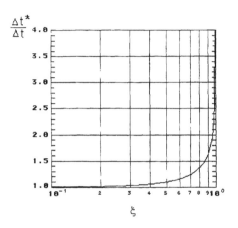

Figure 7.4. *Validity of approximate expression of the time spent between the half-power points*

$$\Delta t^* = \frac{60}{\ln 2}\,\frac{\ln\left(\dfrac{1+1/2\,Q}{1-1/2\,Q}\right)}{R_{om}} \qquad\qquad [7.72]$$

(where Δt^* is expressed in seconds) [SPE 61] [SPE 62] [STE 73]. The number of cycles in this interval is equal to

$$\Delta N^* = f_0\,\Delta t^*$$

$$\Delta N^* = \frac{60}{\ln 2}\,\frac{f_0}{R_{om}}\,\ln\left(\frac{1+1/2\,Q}{1-1/2\,Q}\right) \qquad\qquad [7.73]$$

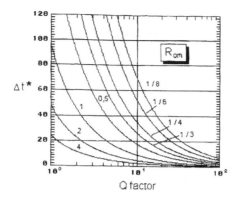

Figure 7.5. *Time spent between the half-power points*

It is noted that Δt^* given by [7.70] tends towards the value given by [7.47] as Q increases. Figure 7.5 shows the variations of Δt^* versus Q for R_{om} (oct/min) equal to 4, 2, 1, 1/2, 1/3, 1/4, 1/6 and 1/8 respectively.

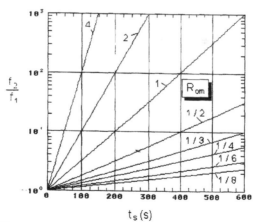

Figure 7.6. *Sweep duration between two frequencies*

Figure 7.6 gives the sweep time necessary to go from f_1 to f_2 for sweep rates R_{om} (oct/min) equal to 4, 2, 1, 1/2, 1/3, 1/4, 1/6 and 1/8.

Number of cycles per octave

If f_A and f_B are two frequencies separated by one octave:

$$f_B = 2 f_A$$

The number of cycles in this octave is equal to [7.41]

$$N_2 = T_1 f_A \qquad\qquad [7.74]$$

i.e., according to [7.42],

$$N_2 = \frac{f_A\, f_b}{\ln 2} \qquad\qquad [7.75]$$

$$N_2 = Q\, \Delta N \frac{f_A}{f_0} \qquad\qquad [7.76]$$

Time to sweep one octave

Let us set t_2 for this duration

$$t_2 = T_1 \ln 2 \qquad\qquad [7.77]$$

$$t_2 = Q\, \Delta t \ln 2 \qquad\qquad [7.78]$$

$$t_2 = \frac{Q\, \Delta N}{f_0} \ln 2 \qquad\qquad [7.79]$$

Time to sweep $1/n^{th}$ octave

$$t_n = T_1\, \ln 2^{1/n} = \frac{T_1}{n} \ln 2 \qquad\qquad [7.80]$$

$$t_n = \frac{Q\, \Delta t}{n} \ln 2 \qquad\qquad [7.81]$$

$$t_n = \frac{Q\, \Delta N}{f_0\, n} \ln 2 \qquad\qquad [7.82]$$

7.4.3. *Case n°2: sweep with constant rate*

If we wish to carry out a sweep with a constant rate, it is necessary that df/dt = constant, i.e., since $\dot{f} = \pm \dfrac{\pi}{\mu\,Q^2} f^2$

$$\mu = \delta\,f^2 \tag{7.83}$$

where δ is a constant with dimension of a time squared.

$$\Delta t = \delta\,f_0^2\,\frac{Q}{f_0}$$

$$\Delta t = \frac{\delta\,f_0\,Q}{\pi} \tag{7.84}$$

.   The time spent in the band Δf delimited by the half-power points varies like the natural frequency f_0.

$$\frac{df}{dt} = \pm \frac{\pi}{Q^2}\frac{f^2}{\delta\,f^2} = \pm \frac{\pi}{Q^2\delta} = \pm\,\alpha \tag{7.85}$$

where α is a constant.

Increasing frequency sweep

$$f = \alpha\,t + f_1 \tag{7.86}$$

The constant α is such that $f = f_2$ when $t = t_s$, yielding [BRO 75], [HOK 48], [LEW 32], [PIM 62], [TUR 54], [WHI 72] and [WHI 82]:

$$\alpha = \frac{f_2 - f_1}{t_b} \tag{7.87}$$

This sweep is known as *linear*.

Decreasing frequency sweep

$$f = -\,\alpha\,t + f_2 \tag{7.88}$$

$$\alpha = \frac{\pi}{Q^2 \delta} = \frac{f_2 - f_1}{t_s} \qquad [7.89]$$

Calculation of the function E(t) [SUN 75]

Increasing frequency:

$$E(t) = 2\pi \int_0^t \left(\alpha t + f_1\right) dt \qquad [7.90]$$

$$E(t) = 2\pi t \left(\frac{\alpha t}{2} + f_1\right) \qquad [7.91]$$

Decreasing frequency:

$$E(t) = 2\pi \int_0^t \left(-\alpha t + f_2\right) dt \qquad [7.92]$$

$$E(t) = 2\pi t \left(-\frac{\alpha t}{2} + f_2\right) \qquad [7.93]$$

Sweep rate

This is equal, depending on the direction of sweep, with

$$\frac{df}{dt} = \pm\alpha = \pm\frac{f_2 - f_1}{t_s} \qquad [7.94]$$

7.4.4. Case n°3: sweep ensuring a number of identical cycles ΔN *in all intervals* Δf *(delimited by the half-power points) for all values of* f_0

With this assumption, since the quantity

$$\Delta N = f\,\Delta t = f\,\frac{\mu\,Q}{\pi\,f} = \mu\,\frac{Q}{\pi} \qquad [7.95]$$

must be constant, the parameter β must be itself constant, yielding:

$$\dot{f} = \pm \frac{\pi \, f^2}{\mu \, Q^2} = \pm a \, f^2 \qquad\qquad [7.96]$$

where $a = \dfrac{\pi}{\mu Q^2}$. The sweep rate varies like the square of the instantaneous frequency. This expression is written [BIC 70] [PAR 61]:

$$\frac{df}{f^2} = \pm a \, t \qquad\qquad [7.97]$$

Increasing frequency sweep between f_1 *and* f_2

By integration,

$$\frac{1}{f_1} - \frac{1}{f} = a \, t \qquad\qquad [7.98]$$

(at $t = 0$, we suppose that $f = f_1$, the starting sweep frequency) i.e. [PAR 61]:

$$f = \frac{f_1}{1 - a f_1 t} \qquad\qquad [7.99]$$

or. since, for $t = t_s$, $f = f_2$:

$$a = \frac{f_2 - f_1}{f_1 f_2 t_s} \qquad\qquad [7.100]$$

In this case, little used in practice, the sweep is known as *hyperbolic* [BRO 75] (also the termed *parabolic sweep*, undoubtedly because of the form of the relation [7.96], and *log-log sweep* [ELD 61] [PAR 61]).

NOTE: *In spite of the form of the denominator of the expression [7.99], the frequency f cannot be negative. For that, it would be necessary that* $1 - a f_1 t < 0$, *i.e.*

$$t > \frac{1}{a f_1} = \frac{t_s f_2}{f_2 - f_1}$$

i.e. that $t > t_s$.

Decreasing frequency sweeping between f_2 and f_1

$$\frac{df}{f_2} = -a\,dt \tag{7.101}$$

$$\frac{1}{f_2} - \frac{1}{f} = -a\,t \tag{7.102}$$

$$f = \frac{f_2}{1 + a\,f_2\,t} \tag{7.103}$$

For $t = t_s$ we have $f = f_1$, yielding:

$$a = \frac{f_2 - f_1}{f_1\,f_2\,t_s} \tag{7.104}$$

Expression for E(t)

The function $E(t)$ in the sine term can be calculated from expression [7.2] of $f(t)$:

$$E(t) = \int_0^t 2\pi\,f(t)\,dt \tag{7.105}$$

– *Increasing frequency sweep [CRU 70] [PAR 61]*

$$E(t) = 2\pi \int_0^t \frac{f_1\,dt}{1 - a\,f_1\,t} = \frac{2\pi}{a} \int_0^{a f_1 t} \frac{d(a\,f_1\,t)}{1 - a\,f_1\,t} \tag{7.106}$$

$$E(t) = -\frac{2\pi}{a} \ln\left(1 - a\,f_1\,t\right)$$

$$E(t) = \frac{2\pi}{a} \ln\left(\frac{1}{1 - a\,f_1\,t}\right) \tag{7.107}$$

i.e. taking into account [7.99]

$$E(t) = \frac{2\pi}{a} \ln \frac{f}{f_1}$$
[7.108]

– *Decreasing frequency sweep*

We have in the same way:

$$E(t) = 2\pi \int_0^t \frac{f_2\, dt}{1 - a\, f_2\, t} = \frac{2\pi}{a} \int_0^{a f_2 t} \frac{d(a\, f_2\, t)}{1 - a\, f_2\, t}$$

$$E(t) = \frac{2\pi}{a} \ln\left(1 + a\, f_2\, t\right)$$
[7.109]

Sweep rate

Increasing frequency:

$$\frac{df}{dt} = a\, f^2 = \frac{f_2 - f_1}{f_1\, f_2\, t_s}\, f^2$$
[7.110]

Decreasing frequency:

$$\frac{df}{dt} = -a\, f^2$$
[7.111]

Tables 8.2 to 8.8 at the end of the Chapter 8 summarize the relations calculated for the three sweep laws (logarithmic, linear and hyperbolic).

Chapter 8

Response of a one-degree-of-freedom linear system to a swept sine vibration

8.1 Influence of sweep rate

An extremely slow sweep rate makes it possible to measure and plot the transfer function of the one-degree-of-freedom system without distortion and to obtain correct values for the resonance frequency and Q factor.

When the sweep rate increases. it is noted that the transfer function obtained differs more and more from the real transfer function. The deformations of the transfer function result in (Figure 8.1):

– a reduction of the maximum δH ;

– a displacement of the abscissa of the maximum δf_r:

– a displacement δf_m of the median axis of the curve (which loses its symmetry);

– an increase in the bandwidth Δf (interval between the half-power points).

When the sweep rate increases:

– one first of all observes on the time history response the appearance of beats which are due to interference between the free response of the mechanical system relatively important after resonance and the excitation 'swept sine' imposed on the system [BAR 48] [PIM 62]. The number and the importance of these beats are weaker since the damping is greater:

– then, as if the system were subjected to a shock, the sweep duration decreases. The largest peak of the response occurs for $t > t_b$ (residual 'response' observed when the duration of the shock is small compared to the natural period of the system). We will see an example of this in paragraph 8.2.3.

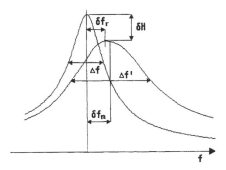

Figure 8.1. *Deformation of the transfer function related to a large sweep rate (according to [REE 60])*

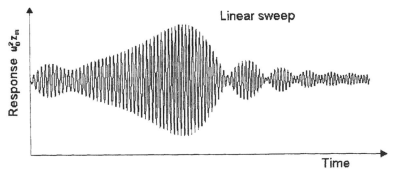

Figure 8.2. *Sweep rate influence on the response of a one-degree-of-freedom system*

8.2. Response of a linear one-degree-of-freedom system to a swept sine excitation

8.2.1. *Methods used for obtaining response*

The calculation of the response of a linear one-degree-of-freedom system cannot be carried out entirely in an analytical way because of the complexity of the equations (except in certain particular cases). Various methods were proposed to solve the differential equation of the movement (analogue [BAR 48], [MOR 65] and [REE 60], numerical [HAW 64]), using the Fourier transformation [WHI 72], the

Laplace transformation [HOK 48], the convolution integral [BAR 48], [LEW 32], [MOR 65], [PAR 61], [SUN 75] and [SUN 80], a series [BAR 48], [MOR 65], [PAR 61], [SUN 75], the Fresnel integrals [DIM 61] [HOK 48], [LEW 32] and [WHI 72], asymptotic developments [KEV 71], techniques of variation of the parameters [SUZ 78a], [SUZ 78b] and [SUZ 79], numerical integration, ...

In general, the transient period of the beginning of the sweep, which relates to only a low number of cycles compared to the total number of cycles of sweep, is neglected. However, it will be preferable to choose the initial frequency of sweep at least an octave below the first resonance frequency of the material to be sure to be completely free from this effect [SUN 80].

8.2.2. Convolution integral (or Duhamel's integral)

We will see in the following paragraphs that the choice of the initial and final frequencies of sweep influence the amplitude of the response, which is all the more sensitive since the sweep rates are larger.

If the excitation is an acceleration the differential equation of the movement of a linear one-degree-of-freedom system is written:

$$m\ddot{z} + c\dot{z} + kz = -m\ddot{x}(t) \qquad [8.1]$$

$$\ddot{z} + 2\xi\,\omega_0\,\dot{z} + \omega_0^2\,z = -\ddot{x}(t) \qquad [8.2]$$

The solution can be expressed in the form of Duhamel's integral:

$$\omega_0^2\,z(t) = -\frac{\omega_0}{\sqrt{1-\xi^2}} \int_0^t \ddot{x}(\lambda)\,e^{\xi\,\omega_0\,(t-\lambda)} \sin\omega_0\sqrt{1-\xi^2}(t-\lambda)\,d\lambda \quad [8.3]$$

if $z(0) = \dot{z}(0) = 0$ (λ = variable of integration). The excitation $\ddot{x}(t)$ is given by:

$$\ddot{x}(t) = \ddot{x}_m\,\sin E(t)$$

where $E(t)$ is given, according to the case involved by [7.91], [7.32] for a logarithmic sweep or by [7.107] for a hyperbolic sweep (increasing frequency). If we set

$$h = \frac{f}{f_0}\left(= \frac{\Omega}{\omega_0}\right)$$

and

$$\theta = \omega_0 \, t$$

these expressions can be written respectively in reduced form:

$$E(\theta) = \theta \left[\frac{h_2 - h_1}{2\,\theta_s} \theta + h_1 \right]$$ [8.4]

$$E(\theta) = \theta_1 \left(h - h_1 \right)$$ [8.5]

$$E(\theta) = - \frac{h_1 \, h_2 \, \theta_s}{h_2 - h_1} \ln \left[1 - \frac{(h_2 - h_1)\theta}{h_2 \, \theta_s} \right]$$ [8.6]

where

$$\theta_1 = \theta_s / \ln(h_2 / h_1)$$

and

$$\theta_s = \omega_0 \, t_s$$

This yields, λ being a variable of integration:

$$q(\theta) = \frac{\omega_0^2 \, z_m}{(-\ddot{x}_m)} = \frac{1}{\sqrt{1 - \xi^2}} \int_0^\theta \sin[E(\lambda)] e^{-\xi(\theta - \lambda)} \sin\left[\sqrt{1 - \xi^2} (\theta - \lambda) \right] d\lambda$$

[8.7]

It is noted that the reduced response $q(\theta)$ is a function of the parameters ξ, θ_s, h_1 and h_2 only and is independent of the natural frequency f_0.

Numerical calculation of Duhamel's integral.

Direct calculation of $q(\theta)$ from numerical integration of [8.7] is possible, but it:

– requires a number of points of integration all larger since the sweep rate is smaller:

– sometimes introduces, for the weak rates, singular points in the plot of $q(\theta)$, which do not necessarily disappear on increasing the number of points of integration (or changing the X-coordinate).

It is in this manner that the results given in the following paragraphs were obtained. Integration was carried out by Simpson's method.

8.2.3. *Response of a linear one-degree-of freedom system to a linear swept sine excitation*

The numerical integration of expression [8.7] was carried out for various values of h_1 and h_2. for $\xi = 0.1$, with between 400 and 600 points of calculation (according to the sweep rate) with, according to the sweep direction, $E(\theta)$ being given by [8.4] if the sweep is at an increasing frequency or by

$$E(\theta) = \theta \left[- \frac{h_2 - h_1}{2\,\theta_s} \theta + h_2 \right] \qquad [8.8]$$

if the frequency is decreasing. On each curve response $q(\theta)$, we have to note:

– the highest maximum:

– the lowest minimum (it was noted that these two peaks always follow each other);

– the frequency of the excitation at the moment when these two peaks occur;

– the frequency of the response around these peaks starting from the relation

$f_R = \dfrac{1}{2\,\Delta\theta}$, with $\Delta\theta$ being the interval of time separating these two consecutive

peaks.

The results are presented in the form of curves in reduced coordinates with:

– on the abscissae, the parameter η defined by:

$$\eta = \frac{Q^2}{f_0^2} \left(\frac{df}{dt} \right)_{f=f_0} \qquad [8.9]$$

We will see that this is used by the majority of authors [BAR 48], [BRO 75], [CRO 56], [CRO 68], [GER 61], [KHA 57], [PIM 62], [SPE 61], [TRU 70], [TRU 95] and [TUR 54], in this form or a very close form $(\frac{7}{\pi}\eta, \frac{2}{\pi}\eta, ...)$.

Since, for a linear sweep and according to the direction of sweep,

$$f = \frac{f_2 - f_1}{t_s} t + f_1 \qquad\qquad [8.10]$$

or

$$f = -\frac{f_2 - f_1}{t_s} t + f_2$$

we have:

$$|\eta| = \frac{Q^2}{f_0^2} \frac{f_2 - f_1}{t_s} \qquad\qquad [8.11]$$

If frequency and time are themselves expressed in reduced form, η can be written:

$$\boxed{\eta = 2\pi Q^2 \left(\frac{dh}{d\theta} \right)_{h=1}} \qquad\qquad [8.12]$$

with for linear sweep with increasing frequency

$$h = \frac{dE}{d\theta} = \frac{h_2 - h_1}{\theta_s} \theta + h_1 \qquad\qquad [8.13]$$

and with decreasing frequency:

$$h = -\frac{h_2 - h_1}{\theta_s} \theta + h_2 \qquad\qquad [8.14]$$

yielding:

$$\frac{dh}{d\theta} = \overset{+}{_-} \frac{h_2 - h_1}{\theta_s} \qquad\qquad [8.15]$$

and

$$|\eta| = 2\pi Q^2 \frac{h_2 - h_1}{\theta_s} \qquad\qquad [8.16]$$

– on the ordinates the ratio G of the largest positive or negative peak (in absolute value) of the response $q(\theta)$ to the largest peak which would be obtained in steady state mode $(Q / \sqrt{1 - \xi^2})$.

Calculations were carried out for sweeps at increasing and decreasing frequency. These showed that:

– for given η results differ according to the values of the limits h_1 and h_2 of the sweep; there is a couple h_1, h_2 for which the peaks of the response are largest. This phenomenon is all the more sensitive since η is larger ($\eta \geq 5$);

– this peak is sometimes positive, sometimes negative;

– for given η, sweep at decreasing frequency leads to responses larger than sweep at increasing frequency.

Figure 8.3 shows the curves $G(\eta)$ thus obtained. These curves are envelopes of all the possible results.

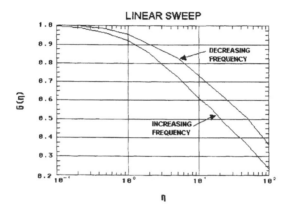

Figure 8.3. *Attenuation of the peak versus the reduced sweep rate*

In addition, Figure 8.4 shows the variations with η of the quantity

$$\frac{Q\,\delta f}{f_R} = \frac{\delta f}{\Delta f}$$

where $\delta f = f_p - f_R$, difference between the peak frequency of the transfer function measured with a fast sweep and resonance frequency $f_R \left(= f_0 \sqrt{1 - 2\,\xi^2} \right)$ measured with a very slow sweep.

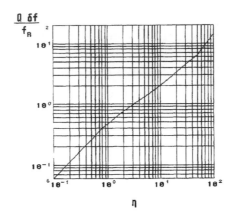

$$\frac{Q}{f_R}\frac{\delta f}{}$$

Figure 8.4. *Shift in the resonance frequency*

The frequency f_p is that of the excitation at the moment when the response passes through the highest peak (absolute value). Δf is the width of the resonance peak measured between the half-power points (with a very slow sweep).

The values of the frequencies selected to plot this curve are those of the peaks (positive or negative) selected to plot the curve $G(\eta)$ of Figure 8.3 (for sweeps at increasing frequency).

NOTE: *These curves have been plotted for η varying between 0.1 and 100. This is a very important domain. To be convincing, it is enough to calculate for various values of η the number of cycles N_b carried out between h_1 and h_2 for given Q. This number of cycles is given by:*

$$N_s = \frac{f_1 + f_2}{2} t_s = \frac{f_1/f_0 + f_2/f_0}{2} \tilde{f_0} \, t_s$$

$$N_s = \frac{h_1 + h_2}{2} \frac{\theta_s}{2\pi} \qquad\qquad [8.17]$$

In addition, we showed [8.16] that

$$\eta = 2\pi Q^2 \left(\frac{h_2 - h_1}{\theta_s} \right)$$

$$\eta = Q^2 \frac{f_2 - f_1}{f_0^2 t_s}$$

$$t_s = Q^2 \frac{f_2 - f_1}{f_0^2 \eta}$$

yielding, since $N_s = \frac{f_1 + f_2}{2} t_s$

$$N_s = \frac{\left(h_2^2 - h_1^2\right) Q^2}{2\eta} \qquad\qquad [8.18]$$

Example

$h_1 = 0.5$
$Q = 5$
$h_2 = 1.5$

If $\eta = 0.1$ there are $N_s = 250$ cycles and if $\eta = 10$, there are $N_s = 2.5$ cycles.

For the higher values of η and for certain couples h_1, h_2, it can happen that the largest peak occurs after the end of sweep ($t > t_s$). There is, in this case, a 'residual' response, the system responding to its natural frequency after an excitation of short duration compared to its natural period ('impulse response'). The swept sine can be considered as a shock.

Example

     $\eta = 60$
     $f_1 = 10$ Hz
     $f_0 = 20$ Hz
     $f_2 = 30$ Hz
     $Q = 5$

With these data, the duration t_b is equal to 20.83 ms.

Figure 8.5 shows the swept sine and the response obtained (velocity: $\dot{f} = 960$ Hz/s).

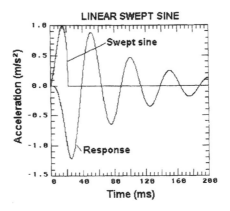

Figure 8.5. *Example of response to a fast swept sine*

It is noted that, for this rate, the excitation resembles a half-sine shock of duration t_s and amplitude 1.

On the shock response spectrum of this half-sine (Figure 8.6), we would read in the ordinate (for $f_0 = 20$ Hz on abscissa) an amplitude of the response of the one-degree-of-freedom system ($f_0 = 20$ Hz, $Q = 5$) equal to 1.22 m/s^2, a value which is that raised above on the curve in Figure 8.5.

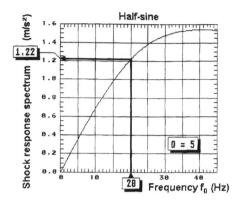

Figure 8.6. *Shock response spectrum of a half-sine shock*

For the same value of η, and with the same mechanical system, we can obtain, by taking $f_1 = 1$ Hz and $f_2 = 43.8$ Hz, an extreme response equal to 1.65 m/s² (Figure 8.7).

In this case the duration has as a value 44.58 ms.

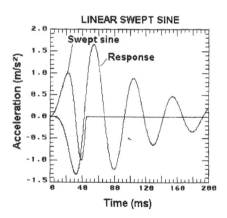

Figure 8.7. *Response for the same η value and for other limit frequencies of the swept domain*

Note on parameter η

As defined by [8.9], this parameter η is none other than the quantity π/μ of the relation [7.22]. If we calculate the number of cycles ΔN according to η carried out in the band Δf delimited by the half-power points, we obtain, according to the sweep mode:

  - linear sweep

$$\Delta N = \frac{f_0^2}{Q} \frac{t_s}{f_2 - f_1}$$

$$\eta = \frac{Q^2}{f_0^2} \frac{f_2 - f_1}{t_s}$$

yielding

$$\Delta N = \frac{Q}{\eta} \qquad\qquad [8.19]$$

  - logarithmic sweep

$$\eta = \frac{Q^2}{f_0 T_1}$$

$$\Delta N = \frac{f_0}{Q} \frac{t_s}{\ln f_2/f_1} = \frac{f_0 T_1}{Q}$$

$$\Delta N = \frac{Q}{\eta} = \frac{Q^2}{f_0 t_s} \ln \frac{f_0}{f_1} \qquad\qquad [8.20]$$

  - hyperbolic sweep

$$\eta = Q^2 \frac{f_2 - f_1}{f_1 f_2 t_s}$$

$$\Delta N = \frac{f_1 f_2 t_s}{Q(f_2 - f_1)}$$

$$\Delta N = \frac{Q}{\eta} \qquad\qquad [8.21]$$

For given Q and η, the number of cycles carried out in the band Δf is thus identical. As a consequence, the time Δt spent in Δf is, whatever the sweep mode, for η and Q constant

$$\Delta t = \frac{Q}{f_0 \eta} \qquad\qquad [8.22]$$

The expressions of the parameters considered in Chapter 7 expressed in terms of η are given in Tables 8.2 to 8.7 at the end of this chapter.

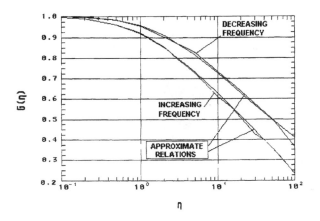

Figure 8.8. *Validity of approximate expressions for attenuation* G(η)

A good approximation of the curve at *increasing frequency* can be obtained by considering the empirical relation (Figure 8.8):

$$G(\eta) = 1 - \exp\left[- 2.55 \ \eta^{-0.39}\right] - 0.003 \ \eta^{0.79} \qquad\qquad [8.23]$$

$(0 \le \eta \le 100)$. To represent the curve G(η) relating to sweeps at *decreasing frequency,* one can use in the same interval the relation:

$$G(\eta) = 1 - \exp\left(- 3.18 \ \eta^{-0.39}\right) \qquad\qquad [8.24]$$

When damping tends towards zero, the time necessary for the establishment of the response tends towards infinity. When the sweep rate is weak, F.M. Lewis [LEW 32] and D.L. Cronin [CRO 68] stated the response of an undamped system as:

$$u_m = 3.67 \sqrt{\dfrac{f_0^2}{\left|\dfrac{df}{dt}\right|_{f = f_0}}} \qquad [8.25]$$

i.e., if sweep is linear, by:

$$u_m = 3.67 \, f_0 \sqrt{\dfrac{t_s}{f_2 - f_1}} \qquad [8.26]$$

NOTE: *For the response of a simple system having its resonance frequency f_0 outside the swept frequency interval (f_1, f_2) in steady state mode, or for an extremely slow sweep, the maximum generalized response is given:*

– for $f_0 < f_1$, by

$$u_m = \dfrac{\ell_m}{\sqrt{\left[1 - \left(\dfrac{f_1}{f_0}\right)^2\right]^2 + \dfrac{f_1^2}{Q^2 \, f_0^2}}} \qquad [8.27]$$

– for $f_0 > f_2$, by

$$u_m = \dfrac{\ell_m}{\sqrt{\left[1 - \left(\dfrac{f_2}{f_0}\right)^2\right]^2 + \dfrac{f_2^2}{Q^2 \, f_0^2}}} \qquad [8.28]$$

When the sweep rate is faster it is possible to obtain an approximate value of the response by successively combining [8.23] and [8.27], [8.23] and [8.28]:

$-\ f_0 < f_1$

$$u_m = \frac{\ell_m \left\{ 1 - \exp\left[-2.55\ \eta^{-0.39} \right] - 0.003\ \eta^{0.79} \right\}}{\sqrt{\left[1 - \left(\dfrac{f_1}{f_0} \right)^2 \right]^2 + \dfrac{f_1^2}{f_0^2\ Q^2}}}$$

[8.29]

$-\ f_0 > f_2$

$$u_m = \frac{\ell_m \left\{ 1 - \exp\left[-2.55\ \eta^{-0.39} \right] - 0.003\ \eta^{0.79} \right\}}{\sqrt{\left[1 - \left(\dfrac{f_2}{f_0} \right)^2 \right]^2 + \dfrac{f_2^2}{f_0^2\ Q^2}}}$$

[8.30]

where η is given by [8.11].

8.2.4. *Response of a linear one-degree-of-freedom system to a logarithmic swept sine*

The calculation of Duhamel's integral [8.7] was carried out under the same conditions as in the case of linear sweep, with:

$$E(\theta) = \theta_1 \left(h - h_1 \right)$$

[8.31]

or

$$E(\theta) = \theta_1 \left(h_2 - h \right)$$

[8.32]

according to the direction of sweep, with $\xi = 0.1$, for various values of the sweep rate, the limits h_1 and h_2 being those which, for each value of η, lead to the largest response (in absolute value). The curves $G(\eta)$ thus obtained were plotted on Figure 8.9, η being equal to:

$$|\eta| = \frac{Q^2}{f_0^2} \left(\frac{df}{dt} \right)_{f=f_0}$$

$$|\eta| = \frac{Q^2}{f_0^2} \left(\frac{f}{T_1} \right)_{f=f_0} = \frac{Q^2}{f_0\ T_1}$$

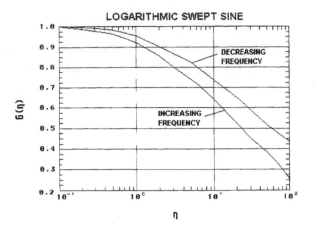

Figure 8.9. *Attenuation versus reduced sweep rate*

$$|\eta| = \frac{2\pi Q^2}{\theta_1} = \frac{2\pi Q^2}{\theta_s}\ln\frac{h_2}{h_1} \qquad\qquad [8.33]$$

where

$$\theta_1 = 2\pi f_0\, T_1 \qquad\qquad [8.34]$$

These curves can be represented by the following empirical relations (for $0 \le \eta \le 100$):

– *for increasing frequencies:*

$$G(\eta) = 1 - \exp\left[-2.55\,\eta^{-0.39}\right] - 0.0025\,\eta^{0.79} \qquad\qquad [8.35]$$

– *for decreasing frequencies:*

$$G(\eta) = 1 - \exp\left[-3.18\,\eta^{-0.38}\right] \qquad\qquad [8.36]$$

Figure 8.10 shows the calculated curves and those corresponding to these relations.

The remarks relating to the curves $G(\eta)$ for the linear sweep case apply completely to the case of logarithmic sweep.

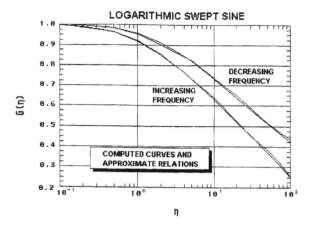

Figure 8.10. *Validity of the approximate expressions for attenuation* G(η)

The number of cycles between h_1 and h_2 is given here by:

$$N_s = \frac{f_2 - f_1}{\ln f_2/f_1} t_s$$

$$N_s = \frac{h_2 - h_1}{\ln h_2/h_1} \frac{\theta_s}{2\pi}$$ [8.37]

yielding, starting from [8.33]:

$$N_s = \frac{Q^2}{\eta} \left(h_2 - h_1 \right)$$ [8.38]

$$\left. \begin{array}{l} \theta_s = \dfrac{2\pi Q^2}{\eta} \ln \dfrac{h_2}{h_1} \\[3mm] t_s = \dfrac{Q^2}{\eta f_0} \ln f_2/f_1 \end{array} \right]$$ [8.39]

Example

Table 8.1. *Examples of sweep durations for given values of* η

If $f_1 = 10$ Hz	η	N_s	t_s (s)
$Q = 5$	0.1	250	137.33
$f_2 = 30$ Hz	10	2.5	0.1373
$f_0 = 20$ Hz	60	0.417	0.02289
	100	0.25	0.01373

Figure 8.11 shows the swept sine (log) for increasing frequency and the response calculated with these data for $\eta = 60$.

It is possible to find other limits of the swept range (f_1, f_2) leading to a larger response.

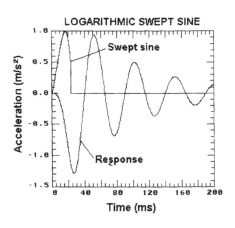

Figure 8.11. *Example of response to a fast swept sine*

The curves $G(\eta)$ obtained in the case of linear and logarithmic sweeps at increasing and decreasing frequencies are superimposed on Figure 8.12. We obtain very similar curves (for a given sweep direction) with these two types of sweeps.

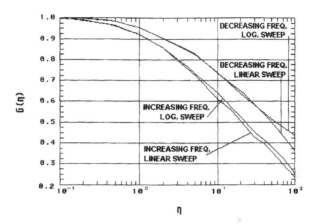

Figure 8.12. *Comparison of the attenuation of linear and logarithmic sweeps*

8.3. Choice of duration of swept sine test

In this paragraph, the duration of the tests intended to simulate a certain particular swept sine real environment will not be considered.

During an identification test intended to measure the transfer function of a mechanical system, it is important to sweep slowly so that the system responds at its resonance with an amplitude very close to the permanent response, whilst adjusting the duration of sweep to avoid prohibitive test times .

It was seen that a good approximation to the measure of the resonance peak can be obtained if η is sufficiently small; J.T. Broch [BRO 75] advised, for example, that $\eta \le 0.1$, which ensures an error lower than $1 - G = 1\%$. For a given $1 - G_0$ error, the curve $G(\eta)$ makes it possible to obtain the limit value η_0 of η not to exceed:

$$\eta = \frac{\left(\dfrac{df}{dt}\right)_{f=f_0} Q^2}{f_0^2} \le \eta_0$$

Table 8.2. *Minimal sweeping duration versus sweeping mode*

Linear sweep	$\dfrac{f_2 - f_1}{t_s} \dfrac{Q^2}{f_0^2} \leq \eta_0$	$t_s \geq \dfrac{f_2 - f_1}{\eta_0 f_0^2} Q^2$
Logarithmic sweep		$t_s \geq \dfrac{Q^2}{f_0 \eta_0} \ln \dfrac{f_2}{f_1}$
	$\dfrac{Q^2}{f_0} \dfrac{\ln(f_2/f_1)}{t_s} \leq \eta_0$	$R_{os} \leq \dfrac{f_0^2 \eta_0 \ln 2}{Q^2}$
		$R_{om} \leq 60 \dfrac{f_0^2 \eta_0 \ln 2}{Q^2}$
Hyperbolic sweep	$Q^2 \dfrac{f_2 - f_1}{f_1 f_2 t_s} \leq \eta_0$	$t_s \geq Q^2 \dfrac{f_2 - f_1}{f_1 f_2 \eta_0}$

It is noted that, in this last case, t_s is independent of f_0. In both other cases, f_0 being in general unknown, one will take for f_0 the value of the swept frequency range which leads to the largest duration t_s .

8.4. Choice of amplitude

To search for resonance frequency the amplitude of the excitation must be:

– sufficiently high to correctly 'reveal' the peaks of resonance in the response. If the structure is linear, the values of Q measured are independent of the sweep level:

– sufficiently weak not to damage the specimen (by exceeding an instantaneous stress level or by fatigue). The choice must thus be made by taking account of the levels of the real vibratory environments.

If the structure is not linear, the value of Q measured depends on the level of the excitation. Generally, Q decreases when the level of the excitation increases. If we wish to use in calculations an experimental transfer function, we will have to measure this function using an excitation in which the level is close to those of the real environments that the structure will exist in. One often has to choose two levels (or more).

8.5. Choice of sweep mode

The more common use of swept sine is for the determination of the dynamic properties of a structure or of a material (natural frequencies, Q factors). For this type of test, the sweep rate must be sufficiently slow so that the response reaches a strong percentage of the response in steady state excitation (it will be seen however (Volume 3) that there are methods using very fast sweeps).

The relations allowing the determination of the test duration are based on the calculations carried out in the case of a linear single-degree-of-freedom system. It is admitted then that if this condition is conformed to, the swept sine thus defined will also create in a several degrees-of-freedom system responses very close to those which one would obtain in steady state mode; this assumption can be criticized for structures having modes with close frequencies.

For this use, it may be worthwhile to choose a sweep mode like that which leads to the weakest duration of test, for the same percentage G of steady state response. According to the relations given in paragraph 8.3, it appears unfortunately that the mode determined according to this criterion is a function of the natural frequency f_0 to be measured in the swept frequency interval. Generally a logarithmic sweep is the preferred choice in practice.

The hyperbolic sweep, little used to our knowledge, presents the property appropriate to carry out a constant number of cycles in each interval Δf delimited by the half-power points of a linear one-degree-of-freedom system, whatever the frequency of resonance (or a constant number of cycles of amplitude higher than P % of the Q factor [CRE 54]).

This property can be used to simulate the effects of a shock (the free response of a system is made with the same number of cycles whatever the resonance frequency f_0, provided that the Q factor is constant) or to carry out fatigue tests. In this case, it must be noted that, if the number of cycles carried out at each resonance is the same, the damage created by fatigue will be the same as if the excitation produces a maximum response displacement z_m (i.e. a stress) identical to all the frequencies.

Gertel [GER 61] advises this test procedure for materials for which the life duration is extremely long (equipment installed on various means of transport, such as road vehicles, aircraft).

The sweep duration t_s can be defined *a priori* or by imposing a given number ΔN of cycles around each resonance, the duration t_s being then calculated from:

$$t_s = Q \, \Delta N \, \frac{f_2 - f_1}{f_1 \, f_2} \qquad\qquad [8.40]$$

by introducing the value of the highest Q factor measured during the identification tests (search for resonance) or a value considered representative in the absence of such tests. The limits f_1 and f_2 delimit a frequency range which must include the principal resonance frequencies of the material. This type of sweep is sometimes used in certain spectral analyzers for the study of experimental signals whose frequency varies with time [BIC 70].

To simulate an environment of short duration Δt, such as the propulsion of a missile, on a material of which the resonance frequencies are little known, it is preferable to carry out a test where each resonance is excited during this time Δt (logarithmic sweep) [PIM 62]. The total duration of sweep is then determined by the relation [7.37]:

$$t_s = Q \, \Delta t \, \ln \frac{f_2}{f_1} \qquad\qquad [8.41]$$

NOTE: *The test duration* t_s *thus calculated can sometimes be relatively long and this, more especially as it is in general necessary to subject to the vibrations the specimen on each of its three axes. Thus, for example, if* $Q = 10$, $\Delta t = 20$ *s,* $f_1 = 10$ *Hz and* $f_2 = 2000$ *Hz*

$$t_s \approx 1060 \ s,$$

yielding a total test duration of 3 x 1060 s = 3180 s (53 min).

C.E. Crede and E.J. Lunney [CRE 56] recommend that a sweep is carried out in several frequency bands simultaneously to save time. The method consists of cutting out the signal (swept sine) which would be held between times t_1 and t_2 in several intervals taken between t_1 and t_a: t_a and t_b, ... , t_n and t_2, and to apply to the specimen the sum of these signals.

Knowing that the material is especially sensitive to the vibrations whose frequency is located between the half-power points, it can be considered that only the swept component which has a frequency near to the frequency of resonance will act notably on the behaviour of the material, the others having little effect. In addition, if the specimen has several resonant frequencies, all will be excited simultaneously in conformity with reality.

Another possibility consists of sweeping the frequency range quickly and in reducing the sweep rate in the frequency bands where the dynamic response is important to measure the peaks correctly.

· *C.F. Lorenzo [LOR 70] proposes a technique of control being based on this principle, usable for linear and logarithmic sweeps, making it possible to reduce with equal precision the test duration by a factor of about 7.5 (for linear sweep).*

The justification for a test with linear sweep clearly does not appear, except if we accept the assumption that the Q factor is not constant whatever the natural frequency f_0. If Q can vary according to a law Q = constant x f_0 (the Q factor being often an increasing function of the natural frequency), it can be shown that the best mode of sweep is the linear sweep.

Table 8.3. *Summary of sweep expressions*

Type of sweep		Hyperbolic	Logarithmic	Linear
Sweep rate		$\dot{f} = \pm \dfrac{\pi}{\mu Q^2} f^2 = \pm a\, f^2$	$\dot{f} = \pm \dfrac{\pi}{\gamma Q^2} f = \pm \dfrac{f}{T_1}$	$\dot{f} = \pm \dfrac{\pi}{\delta Q^2} = \pm \alpha$
Constant η		$\eta = \dfrac{\pi}{\mu} = a\, Q^2$	$\eta = \dfrac{\pi}{\gamma f_0} = \dfrac{Q^2}{T_1 f_0}$	$\eta = \dfrac{\pi}{\delta f_0^2} = \dfrac{\alpha Q^2}{f_0^2}$
		$\eta = Q^2 \dfrac{f_2 - f_1}{f_1 f_2 t_s}$	$\eta = \dfrac{Q^2}{f_0 T_1}$	$\eta = \dfrac{Q^2}{f_0^2} \dfrac{f_2 - f_1}{t_s}$
Law $f(t)$	$f\uparrow$	$f = \dfrac{f_1}{1 - a f_1 t}$	$f = f_1 e^{t/T_1}$	$f = \alpha\, t + f_1$
	$f\downarrow$	$f = \dfrac{f_2}{1 + a f_2 t}$	$f = f_2 e^{-t/T_1}$	$f = -\alpha\, t + f_2$
	$f\uparrow$	$E(t) = -\dfrac{2\pi}{a} \ln(1 - a f_1 t)$	$E = 2\pi T_1 (f - f_1)$	$E = 2\pi t \left(\dfrac{\alpha t}{2} - f_1\right)$
	$f\downarrow$	$E(t) = \dfrac{2\pi}{a} \ln(1 + a f_2 t)$	$E = 2\pi T_1 (f_2 - f)$	$E = 2\pi t \left(-\dfrac{\alpha t}{2} + f_2\right)$
	Cst	$a = \dfrac{f_2 - f_1}{t_s f_1 f_2} = \dfrac{\pi}{\mu Q^2}$	$T_1 = \dfrac{t_b}{\ln(f_2/f_1)} = \dfrac{\gamma Q^2}{\pi}$	$\alpha = \dfrac{\pi}{Q^2 \delta} = \dfrac{f_2 - f_1}{t_s}$
		$a = \dfrac{1}{Q\,\Delta N} = \dfrac{\eta}{Q^2}$	$T_1 = \dfrac{Q^2}{\eta f_0}$	/
η		$Q^2 \dfrac{f_2 - f_1}{f_1 f_2 t_s}$	$\dfrac{Q^2 \ln(f_2/f_1)}{f_0 t_s}$	$\dfrac{Q^2}{f_0^2} \dfrac{f_2 - f_1}{t_s}$

Table 8.4. *Summary of sweep expressions*

Type of sweep	Hyperbolic	Logarithmic	Linear
Number of cycles carried out during t_s Between the frequencies f_1 and f_2	$N_s = \dfrac{1}{a} \ln \dfrac{f_2}{f_1}$	$N_s = T_1 \left(f_2 - f_1 \right)$	$N_s = \dfrac{1}{2a} \left(f_2^2 - f_1^2 \right)$
	$N_s = \dfrac{f_1 f_2}{f_2 - f_1} t_s \ln \dfrac{f_2}{f_1}$	$N_s = \dfrac{f_2 - f_1}{\ln \dfrac{f_2}{f_1}} t_s$	$N_s = \dfrac{f_1 + f_2}{2} t_s$
	$N_s = Q \, \Delta N \ln \dfrac{f_2}{f_1}$	$N_s = Q \, \Delta N \dfrac{f_2 - f_1}{f_0}$	$N_s = \dfrac{Q \, \Delta N}{2 f_0^2} \left(f_2^2 - f_1^2 \right)$
	$N_s = \dfrac{Q^2}{\eta} \ln \dfrac{f_2}{f_1}$	$N_s = \dfrac{Q^2}{\eta f_0} \left(f_2 - f_1 \right)$	$N_s = \dfrac{Q^2}{2 \eta f_0^2} \left(f_2^2 - f_1^2 \right)$
Sweep duration Between the frequencies f_1 and f_2	$t_s = \dfrac{1}{a} \dfrac{f_2 - f_1}{f_1 f_2}$	$t_s = T_1 \ln \dfrac{f_2}{f_1}$	$t_s = \dfrac{1}{\alpha} \left(f_2 - f_1 \right)$
	$t_s = Q \, f_0 \, \Delta t \dfrac{f_2 - f_1}{f_1 f_2}$	$t_s = Q \, \Delta t \ln \dfrac{f_2}{f_1}$	$t_s = \dfrac{Q \, \Delta t}{f_0} \left(f_2 - f_1 \right)$
	$t_s = Q \, \Delta N \dfrac{f_2 - f_1}{f_1 f_2}$	$t_s = \dfrac{Q \, \Delta N}{f_0} \ln \dfrac{f_2}{f_1}$	$t_s = \dfrac{Q \, \Delta N}{f_0^2} \left(f_2 - f_1 \right)$
	$t_s = \dfrac{Q^2}{\eta} \dfrac{f_2 - f_1}{f_1 f_2}$	$t_s = \dfrac{Q^2}{\eta f_0} \ln \dfrac{f_2}{f_1}$	$t_s = \dfrac{Q^2}{\eta f_0^2} \left(f_2 - f_1 \right)$
Interval of time spent in the band Δf	$\Delta t = \dfrac{1}{a \, Q \, f_0}$	$\Delta t = \dfrac{T_1}{Q} = \text{constant}$	$\Delta t = \dfrac{f_0}{\alpha \, Q}$
	$\Delta t = \dfrac{f_1 f_2 t_s}{f_0 \, Q \left(f_2 - f_1 \right)}$	$\Delta t = \dfrac{t_s}{Q \ln f_2 / f_1}$	$\Delta t = \dfrac{f_0 \, t_s}{Q \left(f_2 - f_1 \right)}$
	$\Delta t = \dfrac{Q}{\eta \, f_0}$	$\Delta t = \dfrac{Q}{\eta f_0}$	$\Delta t = \dfrac{Q}{\eta f_0}$

Table 8.5. *Summary of sweep expressions*

Type of sweep	Hyperbolic	Logarithmic curve	Linear
Number of cycles carried out in the interval Δf (between the half-power points) of a one-degree-of-freedom system	$\Delta N = \dfrac{\pi}{a\,Q} = \text{constant}$	$\Delta N = \dfrac{f_0\,T_1}{Q}$	$\Delta N = \dfrac{f_0^2}{\alpha\,Q}$
	$\Delta N = \dfrac{N_s}{\ln\dfrac{f_2}{f_1}}$	$\Delta N = \dfrac{f_0\,N_s}{Q\,(f_2 - f_1)}$	$\Delta N = \dfrac{2\,f_0^2\,N_s}{Q\left(f_2^2 - f_1^2\right)}$
	$\Delta N = \dfrac{f_1\,f_2\,t_s}{Q\,(f_2 - f_1)}$	$\Delta N = \dfrac{f_0\,t_s}{Q\,\ln f_2/f_1}$	$\Delta N = \dfrac{f_0^2\,t_s}{Q\,(f_2 - f_1)}$
	$\Delta N = \dfrac{Q}{\eta}$	$\Delta N = \dfrac{Q}{\eta}$	$\Delta N = \dfrac{Q}{\eta}$
Number of cycles to be carried out between f_1 and f_0 (resonance frequency)	$N_1 = \dfrac{1}{a}\ln\dfrac{f_0}{f_1}$	$N_1 = T_1\left(f_0 - f_1\right)$	$N_1 = \dfrac{1}{2\,\alpha}\left(f_0^2 - f_1^2\right)$
	$N_1 = Q\,\Delta N\,\ln\dfrac{f_0}{f_1}$	$N_1 = \dfrac{Q\,\Delta N}{f_0}\left(f_0 - f_1\right)$	$N_1 = \dfrac{Q\,\Delta N}{2\,f_0^2}\left(f_0^2 - f_1^2\right)$
	$N_1 = \dfrac{f_1\,f_2\,t_s}{f_2 - f_1}\ln\dfrac{f_0}{f_1}$	$N_1 = t_s\dfrac{f_0 - f_1}{\ln f_2/f_1}$	$N_1 = \dfrac{t_s}{2}\dfrac{f_0^2 - f_1^2}{f_2 - f_1}$
	$N_1 = \dfrac{Q^2}{\eta}\ln\dfrac{f_0}{f_1}$	$N_1 = \dfrac{Q^2}{\eta\,f_0}\left(f_0 - f_1\right)$	$N_1 = \dfrac{Q^2}{2\,\eta\,f_0^2}\left(f_0^2 - f_1^2\right)$
	$t_1 = \dfrac{1}{a}\dfrac{f_0 - f_1}{f_0\,f_1}$	$t_1 = T_1\ln\dfrac{f_0}{f_1}$	$t_1 = \dfrac{1}{\alpha}\left(f_0 - f_1\right)$
Time t_1 between f_1 and f_0	$t_1 = Q\,\Delta N\,\dfrac{f_0 - f_1}{f_0\,f_1}$	$t_1 = \dfrac{Q\,\Delta N}{f_0}\ln\dfrac{f_0}{f_1}$	$t_1 = \dfrac{Q\,\Delta N}{f_0^2}\left(f_0 - f_1\right)$

Table 8.6. *Summary of sweep expressions*

Type of sweep	Hyperbolic	Logarithmic curve	Linear
Time t_1 between f_1 and f_0	$t_1 = t_s \dfrac{f_2}{f_0}\dfrac{f_0 - f_1}{f_2 - f_1}$	$t_1 = t_s \dfrac{\ln f_0/f_1}{\ln f_2/f_1}$	$t_1 = t_s \dfrac{f_0 - f_1}{f_2 - f_1}$
	$t_1 = \dfrac{Q^2}{\eta}\dfrac{f_0 - f_1}{f_0 f_1}$	$t_1 = \dfrac{Q^2}{\eta f_0}\ln \dfrac{f_0}{f_1}$	$t_1 = \dfrac{Q^2}{\eta f_0^2}(f_0 - f_1)$
Number of cycles to carry out between $f_1 = 0$ and f_0	/	$N_0 = T_1 f_0$	$N_0 = \dfrac{1}{2\,\alpha} f_0^2$
	/	$N_0 = Q\,\Delta N$	$N_0 = \dfrac{Q\,\Delta N}{2}$
	/	/	$N_0 = \dfrac{f_0^2}{2 f_2} t_s$
	/	/	$N_0 = \dfrac{Q^2}{2\,\eta}$
Time t_0 between $f_1 = 0$ and f_0	/	/	$t_0 = \dfrac{f_0}{\alpha}$
	/	/	$t_0 = \dfrac{Q\,\Delta N}{f_0}$
	/	/	$t_0 = \dfrac{f_0}{f_2} t_s$
	/	/	$t_0 = \dfrac{Q^2}{\eta f_0}$
Mean frequency	$f_m = \dfrac{f_1 f_2}{f_2 - f_1}\ln \dfrac{f_2}{f_1}$	$f_m = \dfrac{f_2 - f_1}{\ln f_2/f_1}$	$f_m = \dfrac{f_1 + f_2}{2}$
Time spent between f_a and $f_c \in (f_1, f_2)$	$t_c - t_a = \left[\dfrac{1}{f_a} - \dfrac{1}{f_c}\right]\dfrac{f_1 f_2}{f_2 - f_1} t_s$	$t_c - t_a = t_s \dfrac{\ln f_c/f_a}{\ln f_2/f_1}$	$t_c - t_a = t_s \dfrac{f_c - f_a}{f_2 - f_1}$
	$t_c - t_a = \dfrac{Q^2}{\eta}\left(\dfrac{1}{f_a} - \dfrac{1}{f_c}\right)$	$t_c - t_a = \dfrac{Q^2}{\eta f_0}\ln \dfrac{f_c}{f_a}$	$t_c - t_a = \dfrac{Q^2}{\eta f_0^2}(f_c - f_a)$

Table 8.7. *Summary of sweep expressions*

Type of sweep	Hyperbolic	Logarithmic curve	Linear
Numbers of cycles per octave (f_A = lower frequency of the octave)	$N_2 = \dfrac{\ln 2}{a}$	$N_2 = T_1\, f_A$	$N_2 = \dfrac{3}{2\,\alpha} f_A^2$
	$N_2 = 2\, f_A\, t_s\, \ln 2$	$N_2 = \dfrac{f_A\, t_s}{\ln 2}$	$N_2 = \dfrac{3\, f_A\, t_s}{2}$
	$N_2 = Q\,\Delta N\, \ln 2$	$N_2 = Q\,\Delta N\, \dfrac{f_A}{f_0}$	$N_2 = \dfrac{3\, Q\,\Delta N}{2\, f_0^2} f_A^2$
	$N_2 = \dfrac{Q^2}{\eta}\ln 2$	$N_2 = \dfrac{3\, Q^2\, f_A^2}{2\,\eta\, f_0^2}$	$N_2 = \dfrac{Q^2\, f_A^2}{\eta\, f_0}$
Time necessary to sweep an octave	$t_2 = \dfrac{1}{2\,a\, f_A}$	$t_2 = T_1\, \ln 2$	$t_2 = \dfrac{f_A}{\alpha}$
	$t_2 = \dfrac{Q\, f_0\, \Delta t}{2\, f_A}$	$t_2 = Q\,\Delta t\, \ln 2$	$t_2 = \dfrac{Q\,\Delta t}{f_0} f_A$
	$t_2 = \dfrac{Q\,\Delta N}{2\, f_A}$	$t_2 = \dfrac{Q\,\Delta N}{f_0}\ln 2$	$t_2 = \dfrac{Q\,\Delta N}{f_0^2} f_A$
	$t_2 = \dfrac{Q^2}{2\,\eta\, f_A}$	$t_2 = \dfrac{Q^2\, \ln 2}{\eta\, f_0}$	$t_2 = \dfrac{Q^2\, f_A}{\eta\, f_0^2}$

Table 8.8. *Summary of sweep expressions*

Type of sweep	Hyperbolic	Logarithmic curve	Linear
Time necessary to sweep $1/n$ th octave	$t_n = \dfrac{1}{a f_A}\, \dfrac{2^{1/n} - 1}{2^{1/n}}$	$t_n = \dfrac{T_1}{n} \ln 2$	$t_n = \dfrac{f_A}{\alpha}\left(2^{1/n} - 1\right)$
	$t_n = \dfrac{Q\, \Delta N}{f_A}\, \dfrac{2^{1/n} - 1}{2^{1/n}}$	$t_n = \dfrac{Q\, \Delta N}{f_0\, n} \ln 2$	$t_n = \dfrac{Q\, \Delta N}{f_0}\, f_A \left(2^{1/n} - 1\right)$
	$t_n = \dfrac{Q f_0\, \Delta t}{f_A}\, \dfrac{2^{1/n} - 1}{2^{1/n}}$	$t_n = \dfrac{Q\, \Delta t}{n} \ln 2$	$t_n = \dfrac{Q\, \Delta t}{f_0}\, f_A \left(2^{1/n} - 1\right)$
	$t_n = \dfrac{Q^2}{\eta f_A}\, \dfrac{2^{1/n} - 1}{2^{1/n}}$	$t_n = \dfrac{Q^2 \ln 2}{\eta f_0\, n}$	$t_n = \dfrac{Q^2}{\eta f_0}\, f_A \left(2^{1/n} - 1\right)$
Sweep rate	/	$R_{om} = \dfrac{60 \ln f_2/f_1}{t_s \ln 2}$	$R = 60\, \dfrac{f_2 - f_1}{t_s}$
	/	$R_{om} = \dfrac{60}{T_1 \ln 2}$	/
		$R_{om} = \dfrac{60\, \eta f_0}{Q^2 \ln 2}$	$R = 60\, \dfrac{\eta f_0^2}{Q^2}$

Appendix

Laplace transformations

A.1. Definition

Consider a real continuous function $f(t)$ of the real definite variable t for all $t \geq 0$ and set

$$F(p) \equiv L[f(t)] = \int_0^\infty e^{-pt} f(t) \, dt \qquad [A.1]$$

(provided that the integral converges). The function $f(t)$ is known as '*original*' or '*object*', the function $F(p)$ as '*image*' or '*transform*'.

Example

Consider a step function applied to $t = 0$ and of amplitude f_m. The integral [A.1] gives simply

$$F(p) = \int_0^\infty e^{-pt} f_m \, dt = f_m \left[-\frac{e^{-pt}}{p} \right]_0^\infty \qquad [A.2]$$

$$F(p) = \frac{f_m}{p} \qquad [A.3]$$

A.2. Properties

In this paragraph some useful properties of this transformation are given, without examples.

A.2.1. *Linearity*

$$L\big[f_1(t) + f_2(t)\big] = L\big[f_1(t)\big] + L\big[f_2(t)\big]$$ [A.4]

$$L\big[c\ f(t)\big] = c\ L\big[f(t)\big]$$ [A.5]

A.2.2. *Shifting theorem (or time displacement theorem)*

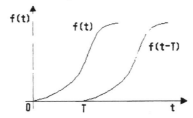

Figure A.1. *Shifting of a curve with respect to the variable* t

Let us set $f(t)$ a transformable function and let us operate a translation parallel to the axis 0t, of amplitude T ($T \geq 0$). If $F(p)$ is the transform of $f(t)$, the transform of $f(t - T)$ is equal to

$$\phi(p) = e^{-p\,T}\ F(p)$$ [A.6]

(formula of the translation on the right or shifting theorem).

Application

A rectangular shock can be considered as being created by the superposition of two levels, one of amplitude f_m *applied at time* $t = 0$ *(transform:* $\dfrac{f_m}{p}$. *cf. preceding example), the other of amplitude* $-f_m$ *applied at time* $t = \tau$, *of transform* $-\dfrac{f_m}{p}\ e^{-p\,\tau}$ *yielding the expression of the transform*

$$L(p) = \frac{f_m}{p} \left(1 - e^{-p\tau}\right)$$ [A.7]

A.2.3. *Complex translation*

$$L\left[f(t)\, e^{-a\,t}\right] = F(p + a)$$ [A.8]

This result makes it possible to write directly the transform of $f(t)\, e^{-a\,t}$ when that of $f(t)$ is known.

A.2.4. *Laplace transform of the derivative of* f(t) *with respect to time*

The transform of the derivative $f'(t)$ of $f(t)$ with respect to t is equal to

$$L[f'(t)] = p\, F(p) - f(0^+)$$ [A.9]

where $F(p)$ is the Laplace transform of $f(t)$ and $f(0^+)$ is the value of the first derivative of $f(t)$ for $t = 0$ (as t approaches zero from the positive side).

In a more general way, the transform of the nth derivative of $f(t)$ is given by

$$\int L\left[\frac{d^n f}{dt^n}\right] = p^n\, F(p) - p^{n-1}\, f(0^+) - p^{n-2}\, f'(0^+) - \cdots - p f^{(n-2)}(0^+) - f^{(n-1)}(0^+)$$ [A.10]

where $f^{(n-1)}(0^+),\ f^{(n-2)}(0^+),\ ....\ f'(0^+)$ are the successive derivatives of $f(t)$ for $t = 0$ (as t approaches zero from the positive side).

A.2.5. *Derivative in the* p *domain*

The nth derivative of the transform $F(p)$ of a function $f(t)$ with respect to the variable p is given by

$$\frac{d^n F}{dp^n} = (-1)^n\, L\left[t^n\, f(t)\right]$$ [A.11]

A.2.6. *Laplace transform of the integral of a function* f(t) *with respect to time*

If $\lim \int_0^\varepsilon f(t) \, dt = 0$ when $\varepsilon \to 0$,

$$L\left[\int_0^t f(t) \, dt\right] = \frac{F(p)}{p} \tag{A.12}$$

and to the order n

$$L\left[\int_0^t dt \int_0^t dt \cdots \int_0^t f(t) \, dt\right] = \frac{F(p)}{p^n} \tag{A.13}$$

A.2.7. *Integral of the transform* $F(p)$

The inverse transform of the integral of $F(p)$ between p and infinite is equal to:

$$\int_p^\infty F(p) \, dp = L\left[\frac{f(t)}{t}\right] \tag{A.14}$$

When integrating n times between the same limits, it becomes

$$\int_p^\infty dp \int_p^\infty dp \cdots \int_p^\infty F(p) \, dp = L\left[\frac{f(t)}{t^n}\right] \tag{A.15}$$

A.2.8. *Scaling theorem*

If a is a constant,

$$L\left[f\left(\frac{t}{a}\right)\right] = a \, F(a \, p) \tag{A.16}$$

$$L[f(a \, t)] = \frac{1}{a} F\left(\frac{p}{a}\right) \tag{A.17}$$

A.2.9. *Theorem of damping or rule of attenuation*

$$F(p + a) = \int_0^\infty e^{-p t} e^{-a t} f(t) \, dt \qquad\qquad [A.18]$$

The inverse transform of $F(p + a)$ is thus $e^{-a t} f(t)$. It is said that the function $e^{-a t}$ damps the function $f(t)$ when a is a positive real constant.

A.3. Application of Laplace transformation to the resolution of linear differential equations

In the principal use made of Laplace transformation, the interest resides in the property relative to derivatives and integrals which becomes, after transformation, products or quotients of the transform $F(p)$ of $f(t)$ by p or its powers.

Let us consider, for example, the second order differential equation

$$\frac{d^2 q(t)}{dt^2} + a \frac{dq(t)}{dt} + b \, q(t) = f(t) \qquad\qquad [A.19]$$

where a and b are constants. Let us set $Q(p)$ and $F(p)$ the Laplace transforms of $q(t)$ and $f(t)$ respectively. From the relations of the paragraph A.2.4, we have

$$L[\ddot{q}(t)] = p^2 \, Q(p) - p \, q(0) - \dot{q}(0) \qquad\qquad [A.20]$$

$$L[\dot{q}(t)] = p \, Q(p) - q(0) \qquad\qquad [A.21]$$

where $q(0)$ and $\dot{q}(0)$ are the values of $q(t)$ and its derivative for $t = 0$. Because of the linearity of the Laplace transformation, it is possible to transform each member of the differential equation [A.19] term by term:

$$L[\ddot{q}(t)] + a \, L[\dot{q}(t)] + b \, L[q(t)] = L[f(t)] \qquad\qquad [A.22]$$

by replacing each transform by its expression this becomes

$$p^2 \, Q(p) - p \, q(0) - \dot{q}(0) + a \left[p \, Q(p) - q(0) \right] + b \, Q(p) = F(p) \qquad [A.23]$$

$$Q(p) = \frac{F(p) + p \, q(0) + a \, q(0) + \dot{q}(0)}{p^2 + a \, p + b} \qquad\qquad [A.24]$$

Let us expand the rational fraction $\dfrac{F(p) + p\,q(0) + a\,q(0) + \dot{q}(0)}{p^2 + ap + b}$ into partial fractions; while noting that p_1 and p_2 are the roots of the denominator $p^2 + ap + b$, we have

$$\frac{A\,p + B}{p^2 + a\,p + b} = \frac{C}{p - p_1} + \frac{D}{p - p_2} \qquad [A.25]$$

with

$$A = q(0) \qquad\qquad C = \frac{A\,p_1 + B}{p_1 - p_2}$$

$$B = a\,q(0) + \dot{q}(0) \qquad\qquad D = -\frac{A\,p_2 + B}{p_1 - p_2}$$

yielding

$$Q(p) = \frac{F(p)}{p^2 + a\,p + b} + \frac{1}{p_1 - p_2}\left[\frac{A\,p_1 + B}{p - p_1} - \frac{A\,p_2 + B}{p - p_2}\right] \qquad [A.26]$$

i.e.

$$Q(p) = \frac{F(p)}{p_1 - p_2}\left[\frac{1}{p - p_1} - \frac{1}{p - p_2}\right]$$

$$+ \frac{1}{p_1 - p_2}\left[\frac{q(0)\,p_1 + a\,q(0) + \dot{q}(0)}{p - p_1} - \frac{q(0)\,p_2 + a\,q(0) + \dot{q}(0)}{p - p_2}\right]$$

$$\qquad\qquad\qquad\qquad [A.27]$$

$$Q(p) = \int_0^t \frac{f(\lambda)}{p_1 - p_2}\left[e^{p_1\,(t-\lambda)} - e^{p_2\,(t-\lambda)}\right]d\lambda$$

$$+ \frac{1}{p_1 - p_2}\left\{[q(0)\,p_1 + a\,q(0) + \dot{q}(0)]\,e^{p_1\,t} - [q(0)\,p_2 + a\,q(0) + \dot{q}(0)]\,e^{p_2\,t}\right\}$$

$$\qquad\qquad\qquad\qquad [A.28]$$

where λ is a variable of integration. In the case of a system initially at rest, $q(0) = \dot{q}(0) = 0$ and

$$Q(p) = \int_0^t \frac{f(\lambda)}{p_1 - p_2} \left[e^{p_1 (t-\lambda)} - e^{p_2 (t-\lambda)} \right] d\lambda \qquad\qquad [A.29]$$

A.4. Calculation of inverse transform: Mellin–Fourier integral or Bromwich transform

Once the calculations are carried out in the domain of p, where they are easier, it is necessary to return to the time domain and to express the output variables as a function of t.

We saw that the Laplace transform $F(p)$ of a function $f(t)$ is given by [A.1]

$$F(p) \equiv L[f(t)] = \int_0^\infty e^{-p t} f(t)\, dt \qquad\qquad [A.30]$$

The inverse transformation is defined by the integral known as the Mellin-Fourier integral

$$L^{-1}[F(p)] \equiv f(t) = \frac{1}{2\pi i} \int_{C-i\infty}^{C+i\infty} F(p)\, e^{p t}\, dp \qquad\qquad [A.31]$$

and calculated, for example, on a Bromwich contour composed of a straight line parallel to the imaginary axis, of positive abscissea C [ANG 61], C being such that all the singularities of the function $F(p)\, e^{p t}$ are on the left of the line [BRO 53]: this contour thus goes from $C - i\infty$ to $C + i\infty$.

If the function $F(p)\, e^{p t}$ has only poles, then the integral is equal to the sum of the corresponding residues, multiplied by $2\pi i$. If this function has singularities other than poles, it is necessary to find, in each case, a equivalent contour to the Bromwich contour allowing calculation of the integral [BRO 53] [QUE 65].

The two integrals [A.1] and [A.31] establish a one-to-one relationship between the functions of t and those of p.

These calculations can be in practice rather heavy and it is prefered to use when possible tables of inverse transform providing the transforms of the most current functions directly [ANG 61] [DIT 67], [HLA 69] and [SAL 71]. The inverse transformation is also done using these tables after having expressed results as a function of p in a form revealing transforms whose inverse transform appears in Table A.1.

Example

Let us consider the expression of the response of a one-degree-of-freedom damped system subjected to a rectangular shock of amplitude unit of the form $f(t) = 1$ and of duration τ. For this length of time τ, i.e. for $t \leq \tau$, the Laplace transform is given by (Table A.1):

$$F(p) = \frac{1}{p} \qquad\qquad\text{[A.32]}$$

NOTE: *After the end of the shock, it would be necessary to use the relation* $F(p) = \dfrac{1 - e^{-p\tau}}{p}.$

Relation [A.24] applies with $a = 2\,\xi$ and $b = 1$, yielding

$$q(t) = L^{-1}\left[\frac{\dfrac{1}{p} + p\,q_0 + 2\,\xi\,q_0 + \dot{q}_0}{\left(p^2 + 2\,\xi\,p + 1\right)}\right] \qquad\qquad\text{[A.33]}$$

$$q(t) = L^{-1}\left[\frac{1}{p\left(p^2 + 2\,\xi\,p + 1\right)}\right] + q_0\,L^{-1}\left[\frac{p}{p^2 + 2\,\xi\,p + 1}\right]$$

$$+ \left(2\,\xi\,q_0 + \dot{q}_0\right) L^{-1}\left[\frac{1}{p^2 + 2\,\xi\,p + 1}\right] \qquad\qquad\text{[A.34]}$$

yielding, using Table A.1 ($\xi \neq 1$):

$$q(t) = 1 - \frac{e^{-\xi t}}{\sqrt{1-\xi^2}} \left(\xi \sin \sqrt{1-\xi^2} \; t + \sqrt{1-\xi^2} \; \cos \sqrt{1-\xi^2} \; t \right)$$

$$+ q_0 \frac{e^{-\xi t}}{\sqrt{1-\xi^2}} \left(\sqrt{1-\xi^2} \; \cos \sqrt{1-\xi^2} \; t - \xi \sin \sqrt{1-\xi^2} \; t \right)$$

$$+ \left(2 \xi q_0 + \dot{q}_0 \right) \frac{e^{-\xi t}}{\sqrt{1-\xi^2}} \sin \sqrt{1-\xi^2} t \qquad \text{[A.35]}$$

$$q(t) = 1 + \frac{e^{-\xi t}}{\sqrt{1-\xi^2}} \left\{ \sqrt{1-\xi^2} \left(q_0 - 1 \right) \cos \sqrt{1-\xi^2} \; t - \left[\xi \left(1 - q_0 \right) - \dot{q}_0 \right] \sin \sqrt{1-\xi^2} \; t \right\}$$

$$\text{[A.36]}$$

A.5. Laplace transforms

Table A.1. *Laplace transforms*

Function $f(t)$	Transform $L[f(t)] = F(p)$
1	$\dfrac{1}{p}$
t	$\dfrac{1}{p^2}$
e^{at}	$\dfrac{1}{p-a}$
$\sin a t$	$\dfrac{a}{p^2 + a^2}$
$\cos a t$	$\dfrac{p}{p^2 + a^2}$
$\text{sh } a t$	$\dfrac{a}{p^2 - a^2}$
$\text{ch } a t$	$\dfrac{p}{p^2 - a^2}$

t^2	$\dfrac{2}{p^3}$
t^n	$\dfrac{n!}{p^{n+1}}$ (n integer ≥ 0)
$\sin^2 t$	$\dfrac{2}{p\left(p^2+4\right)}$
$\cos^2 t$	$\dfrac{p^2+2}{p\left(p^2+4\right)}$
$a\,t - \sin a\,t$	$\dfrac{a^3}{p^2\left(p^2+a^2\right)}$
$\sin a\,t - a\,t\cos a\,t$	$\dfrac{2\,a^3}{\left(p^2+a^2\right)^2}$
$t \sin a\,t$	$\dfrac{2\,a\,p}{\left(p^2+a^2\right)^2}$
$\sin a\,t + a\,t\cos a\,t$	$\dfrac{2\,a\,p^2}{\left(p^2+a^2\right)^2}$
$t \cos a\,t$	$\dfrac{p^2-a^2}{\left(p^2+a^2\right)^2}$
$a \sin b\,t - b \sin a\,t$	$\dfrac{a\,b\left(a^2-b^2\right)}{\left(p^2+a^2\right)\left(p^2+b^2\right)}$
$\dfrac{1}{2\,a^3}\left(\sin a\,t - a\,t\cos a\,t\right)$	$\dfrac{1}{\left(p^2+a^2\right)^2}$
$\dfrac{\cos a\,t - \cos b\,t}{b^2-a^2}$	$\dfrac{p}{\left(p^2+a^2\right)\left(p^2+b^2\right)}$
$\dfrac{e^{-a\,t}-e^{-b\,t}}{b-a}$	$\dfrac{1}{\left(p+a\right)\left(p+b\right)}$

$\dfrac{b\,e^{-bt} - a\,e^{-at}}{b - a}$	$\dfrac{p}{(p+a)(p+b)}$
$t\,e^{at}$	$\dfrac{1}{(p-a)^2}$
$t^n\,e^{at}$	$\dfrac{n!}{(p-a)^{n+1}}\,(n = 1, 2, 3, ...)$
$e^{-at}\cos b\,t$	$\dfrac{p+a}{(p+a)^2 + b^2}$
$e^{-at}\sin b\,t$	$\dfrac{b}{(p+a)^2 + b^2}$
$1 - \dfrac{e^{-\frac{at}{2}}}{\sqrt{1 - \frac{a^2}{4}}}\left[\dfrac{a}{2}\sin\sqrt{1 - \dfrac{a^2}{4}}\,t + \sqrt{1 - \dfrac{a^2}{4}}\cos\sqrt{1 - \dfrac{a^2}{4}}\,t\right]$	$\dfrac{1}{p\left(p^2 + a\,p + 1\right)}$
$\dfrac{e^{-\xi t}}{\sqrt{1 - \xi^2}}\left[\sqrt{1 - \xi^2}\cos\sqrt{1 - \xi^2}\,t - \xi\sin\sqrt{1 - \xi^2}\,t\right]$ $(\xi < 1)$	$\dfrac{p}{p^2 + 2\xi\,p + 1}$
$\dfrac{e^{-\xi t}}{\sqrt{1 - \xi^2}}\sin\sqrt{1 - \xi^2}\,t$	$\dfrac{1}{p^2 + 2\xi\,p + 1}$
$\dfrac{e^{-\xi h t}}{h\sqrt{1 - \xi^2}}\sin h\sqrt{1 - \xi^2}\,t$	$\dfrac{1}{p^2 + 2h\xi\,p + h^2}$
$e^{-\xi h t}\cos h\sqrt{1 - \xi^2}\,t - \dfrac{\xi\,e^{-\xi h t}}{\sqrt{1 - \xi^2}}\sin h\sqrt{1 - \xi^2}\,t$	$\dfrac{p}{p^2 + 2h\xi\,p + h^2}$
$\dfrac{e^{-\xi t}\left(\sin\sqrt{1 - \xi^2}\,t - t\sqrt{1 - \xi^2}\cos\sqrt{1 - \xi^2}\,t\right)}{2\left(1 - \xi^2\right)^{3/2}}$	$\dfrac{1}{\left(p^2 + 2\xi\,p + 1\right)^2}$
$\dfrac{t\,e^{-\xi t}}{2\sqrt{1 - \xi^2}}\sin\sqrt{1 - \xi^2}\,t$	$\dfrac{p}{\left(p^2 + 2\xi\,p + 1\right)^2}$

$t\,e^{-\xi\,\Omega\,t}\,\sin\Omega\,t$	$-2\,\dfrac{(p+\xi\,\Omega)\,\Omega}{\left[(p+\xi\,\Omega)^2+\Omega^2\right]^2}$
$t\,e^{-\xi\,\Omega\,t}\,\cos\Omega\,t$	$\dfrac{p^2+2\,\xi\,\Omega\,p+\xi^2\left(\Omega^2-1\right)}{\left[(p+\xi\,\Omega)^2+\Omega^2\right]^2}$
$\dfrac{1+\xi^2}{2}\,e^{-\xi\,\Omega\,t}\left[\dfrac{\sin\Omega\,t}{\Omega}-t\,\cos(\Omega\,t+\phi)\right]$	$\dfrac{p^2}{\left[(p+\xi\,\Omega)^2+\Omega^2\right]^2}$
$\dfrac{1+\xi^2}{2}\,e^{-\xi\,\Omega\,t}\left[\dfrac{\sin\Omega\,t}{\Omega}-t\,\cos(\Omega\,t+\phi)\right]$ $-t\,e^{-\xi\,\Omega\,t}\,\cos\Omega\,t-\xi\,t\,e^{-\xi\,\Omega\,t}\,\sin\Omega\,t$	$\dfrac{\Omega^2\left(1+\xi^2\right)}{\left[(p+\xi\,\Omega)^2+\Omega^2\right]^2}$
$t-2\,\xi+e^{-\xi\,t}\left[2\,\xi\,\cos\sqrt{1-\xi^2}\,t+\dfrac{2\,\xi^2-1}{\sqrt{1-\xi^2}}\,\sin\sqrt{1-\xi^2}\,t\right]$	$\dfrac{1}{p^2\left(p^2+2\,\xi\,p+1\right)}$

These transforms can be used to calculate others, for example starting from decompositions in partial fractions such as

$$\frac{1}{p\left(p^2+a\,p+1\right)}=\frac{1}{p}-\frac{p+a}{p^2+a\,p+1} \qquad\text{[A.37]}$$

$$\frac{1}{p^2\left(p^2+a\,p+1\right)}=\frac{1}{p^2}-\frac{a}{p}+\frac{a\,p}{p^2+a\,p+1}+\frac{a^2-1}{p^2+a\,p+1} \qquad\text{[A.38]}$$

where, in these relations, there arises, $a=2\,\xi$.

A.6. Generalized impedance – the transfer function

If the initial conditions are zero, the relation [A.24] can be written

$$\left(p^2+a\,p+b\right)Q(p)=F(p) \qquad\text{[A.39]}$$

i.e., while setting

$$Z(p)=p^2+a\,p+b \qquad\text{[A.40]}$$

$$F(p) = Z(p) Q(p)$$ [A.41]

By analogy with the equation which links together the current $I(\Omega)$ (output variable) and the tension $E(\Omega)$ (input variable) in an electrical supply network in sinusoidal mode

$$E(\Omega) = Z(\Omega) I(\Omega)$$ [A.42]

$Z(p)$ is called the *generalized impedance of the system,* and $Z(\Omega)$ is the *transfer impedance* of the circuit. The inverse of $A(p)$ of $Z(p)$, $1/Z(p)$, is the *operational admittance.* The function $A(p)$ is also termed the *transfer function.* It is by its intermediary that the output is expressed versus the input:

$$Q(p) = \frac{1}{Z(p)} F(p) = A(p) F(p)$$ [A.43]

Vibration tests: a brief historical background

Vibration tests on aircraft were developed from 1940 to verify the resistance of parts and equipment prior to their first use [BRO 67].

Such tests became necessary as a result of:

– the increasing complexity of on-board flight equipment which was more sensitive to vibrations;

– improved performance of aircraft (and, more generally, of vehicles), to the extent that the sources of vibration initially localized in engines became extended substantially outwards to the ambient medium (aerodynamic flows).

The chronology of such developments can be summarized as follows [PUS 77]:

1940   Measurement of resonance frequencies.
Self-damping tests.
Sine tests (at fixed frequency) corresponding to the frequencies created by engines running at a permanent speed.
Combined tests (temperature, humidity, altitude).

1946   First electrodynamic exciter [DEV 47] [IMP 47].

1950   Swept sine tests to simulate the variations in engine speed or to excite all the resonance frequencies of the test item of whatever value.

1953   Specifications and tests with random vibrations (introduction of jet engines, simulation of jet flows and aerodynamic turbulences with continuous spectra). These tests were highly controversial until the 1960's [MOR 53]. To overcome the insufficient power of such installations, attempts were made to promote swept narrow-band random variations in the frequency domain of interest [OLS 57].

1955   First publications on acoustic vibrations (development of jet rockets and engines, effect of acoustic vibrations on their structures and equipment).

1957   First acoustic chambers [BAR 57] [COL 59] [FRl 59].

1960   The specification of random vibration became essential and the determination of equivalence between random and sinusoidal vibrations.

1965   J.W. Cooley and J.W. Tukey's algorithm for calculating FFTs [COO 65].

1967   Increasing number of publications on acoustic vibrations.

1970   Tri-axial test facility [DEC 70].
Development of digital control systems.

1976   Extreme response spectra and fatigue damage spectra developed; useful in writing specifications (a method in four stages starting from the life cycle profile).

1984   Account taken of the tailoring of tests in certain standard documents (MIL-[STD 810 D], [GAM 92] (development of specifications on the basis of measuring the real environment).

1995   Customization of the product tailored to its environment. Account taken of the environment throughout all the phases of the project (according to the R.G. Aero 00040 Recommendation).

Bibliography

[AER 62]   'Establishment of the approach to, and development of, interim design criteria for sonic fatigue', Aeronautical Systems Division, Flight Dynamics Laboratory, Wright–Patterson Air Force Base, Ohio, *ASD-TDR 62-26, AD 284597*, June 1962.

[AKA 69]   AKAIKE H. and SWANSON S.R., 'Load history effects in structural fatigue', *Proceedings of the 1969 Annual Meeting IES*, April 1969.

[ANG 61]   ANGOT A., 'Compléments de mathématiques', *Editions de la Revue d'Optique*, Collection Scientifique et Technique du CNET, 1961.

[BAN 77]   BANDSTRA J.P., 'Comparison of equivalent viscous damping and nonlinear damping in discrete and continuous vibrating systems', *Masters Thesis*, University of Pittsburgh, 1977 or *Transactions of the ASME*, Vol. 15, July 1983, p. 382/392.

[BAR 48]   BARBER N.F. and URSELL F., 'The response of a resonant system to a gliding tone', *Philosophical Magazine*, Series 7, Vol. 39, 1948, p. 354/361.

[BAR 57]   BARUCH J.J., 'A new high-intensity noise-testing facility', *The Shock and Vibration Bulletin*, n° 25, Part II, Dec. 1957, p. 25/30.

[BAR 61]   BARTON M.V., CHOBOTOV V. and FUNG Y.C., *A Collection of Information on Shock Spectrum of a Linear System*, Space Technology Laboratories, July 1961.

[BAS 75]   BASTENAIRE F., 'Estimation et prévision statistiques de la résistance et de la durée de vie des matériaux en fatigue', *Journées d'Etude sur la Fatigue*, Université de Bordeaux I, 29 May 1975.

[BEA 80]   BEARDS C.F., 'The control of structural vibration by frictional damping in joints', *Journal of the Society of Environmental Engineers*, Vol. 19, n° 2 (85), June 1980, p. 23/27.

[BEA 82]   BEARDS C.F., 'Damping in structural joints', *The Shock and Vibration Digest*, Vol. 14, n° 6, June 1982, p. 9/11.

[BEN 62]   BENDAT J.S., ENOCHSON L.D., KLEIN G.H. and PIERSOL A.G., 'Advanced concepts of stochastic processes and statistics for flight vehicle vibration estimation and measurement', *ASD-TDR-62-973*, Dec. 1962.

[BEN 63]   BENDAT J.S., ENOCHSON L.D. and PIERSOL A.G., 'Analytical study of vibration data reduction methods', The Technical Products Company, Los Angeles, Contract NAS8-5093, Sept. 1963.

[BEN 71]   BENDAT J.S. and PIERSOL A.G., Random Data: Analysis and measurement procedures, Wiley Interscience, 1971.

[BER 73]   BERT C.W., 'Material damping: an introductory review of mathematical models, measures and experimental techniques', Journal of Sound and Vibration. Vol. 29, n° 2, 1973, p. 129/153.

[BER 76]   BERGMAN L.A. and HANNIBAL A.J., 'An alternate approach to modal damping as applied to seismic-sensitive equipment', The Shock and Vibration Bulletin. n° 46, Part 2, Aug. 1976, p. 69/83.

[BIC 70]   BICKEL H.J. and CITRIN A., 'Constant percentage bandwidth analysis of swept–sinewave data', Proceedings of the IES, 1970, p. 272/276.

[BIR 77]   BIRCHAK J.R.. 'Damping capacity of structural materials', The Shock and Vibration Digest, Vol. 9, n° 4, April 1977, p. 3/11

[BIS 55]   BISHOP R.E.D., 'The treatment of damping forces in vibration theory', Journal of the Royal Aeronautical Society, Vol. 59, Nov. 1955, p. 738.

[BLA 56]   BLAKE R.E., BELSHEIM R.O. and WALSH J.P., 'Damaging potential of shock and vibration', ASME Publication – Shock and Vibration Instrumentation, 1956, p. 147/163.

[BLA 61]   BLAKE R.E., 'Basic vibration theory', in HARRIS C.M. and CREDE C.E. (Eds), Shock and Vibration Handbook, Mc Graw-Hill Book Company, Inc., Vol. 1, n° 2, 1961, p. 1/27.

[BRO]   BROCH J.T., Mechanical Vibration and Shock Measurements, Brüel and Kjaer, Denmark, Naerum.

[BRO 53]   BRONWELL A., Advanced Mathematics in Physics and Engineering, McGraw-Hill Book Company, Inc., 1953.

[BRO 62]   BROWN D., 'Digital techniques for shock and vibration analysis', Society of Automotive Engineers, 585E, National Aerospace Engineering and Manufacturing Meeting, Los Angeles, Calif., 8–12 Oct. 1962.

[BRO 67]   BROCH J.T., 'Essais en vibrations. Les raisons et les moyens', Bruël and Kjaer, Technical Review, n° 3, 1967.

[BRO 75]   BROCH J.T., 'Sur la mesure des fonctions de réponse en fréquence', Bruël and Kjaer, Technical Review, n° 4, 1975.

[BUR 59]   BURGESS J.C., Quick estimation of damping from free damped oscillograms, WADC TR 59 - 676.

[BYE 67]   BYERS J.F., 'Effects of several types of damping on the dynamical behavior of harmonically forced single-degree-of-freedom systems', DRL Acoustical Report n° 272 - AD.A0 36696/5GA, 2 Jan. 1967.

[CAM 53]   CAMPBELL J.D., 'An investigation of the plastic behavior of metal rods subjected to longitudinal impact', Journal of Mechanics and Physics of Solids, Vol. 1, 1953, p. 113.

[CAP 82]   CAPRA A. and DAVIDOVICI V., Calcul Dynamique des Structures en Zones Sismique, Eyrolles, 1982.

[CAU 59]   CAUGHEY T.K., 'Response of a nonlinear string to random loading', *Journal of Applied Mechanics, Transactions of the ASME*, 26 Sept. 1959, p. 341/344.

[CHE 66]   CHENG D.K., *Analysis of Linear Systems*, Addison Wesley Publishing Company, Inc., 1966.

[CLA 49]   CLARK D.S. and WOOD D.S., 'The time delay for the initiation of plastic deformation at rapidly applied constant stress', *Proceedings of the American Society for Testing Materials*, Vol. 49, 1949, p. 717/735.

[CLA 54]   CLARK D.S., 'The behaviour of metals under dynamic loading', *Transactions of the American Society for Metals*, Vol. 46, 1954, p. 34/62.

[CLO 80]   CLOUGH R.W. and PENZIEN J., *Dynamique des Structures, Vol. 1· Principes fondamentaux*, Editions Pluralis, 1980.

[COL 59]   COLE J.H., VON GIERKE H.E., OESTREICHER H.L. and POWER R.G., 'Simulation of random acoustic environments by a wide bande noise siren', *The Shock and Vibration Bulletin*, n° 27, Part II, June 1959, p. 159/168.

[COO 65]   COOLEY J.W. and TUKEY J.W., 'An algorithm for the machine calculation of complex Fourier series', *Mathematics of Computation*, Vol. 19, April 1965. p. 297/301

[CRA 58]   CRANDALL S.H., *Random Vibration*, The M.I.T. Press, Massachussetts Institute of Technology, Cambridge, Massachussets, 1958.

[CRA 62]   CRANDALL S.H., 'On scaling laws for material damping', *NASA-TND-1467*, Dec. 1962.

[CRE 54]   CREDE C.E., GERTEL M. and CAVANAUGH R.D., 'Establishing vibration and shock tests for airborne electronic equipment', *WADC Technical Report n° 54 - 272*, June 1954.

[CRE 56]   CREDE C.E. and LUNNEY E.J., 'Establishment of vibration and shock tests for missile electronics as derived from the measured environment', *WADC Technical Report n° 56 - 503. ASTIA Document n° AD 118133*, 1 Dec. 1956.

[CRE 61]   CREDE C.E. and RUZICKA J.E., *Theory of Vibration Isolation. Shock and Vibration Handbook*, McGraw-Hill Book Company, Vol. 2, p. 30, 1961.

[CRE 65]   CREDE C.E., *Shock and Vibration Concepts in Engineering Design*, Prentice-Hall, Inc., Englewood Cliffs, NJ. 1965

[CRO 56]   CRONIN D.L., 'Response of linear viscous damped systems to excitations having time-varying frequency', *Calif. Instit. Technol-Dynam-Lab Rept*, 1956.

[CRO 68]   CRONIN D.L., 'Response spectra for sweeping sinusoidal excitations', *The Shock and Vibration Bulletin*, n° 38, Part 1, Aug. 1968, p. 133/139

[CRU 70]   CRUM J.D. and GRANT R.L., 'Transient pulse development', *The Shock and Vibration Bulletin*, Vol. 41, Part 5, Dec. 1970, p. 167/176.

[CUR 55]   CURTIS A.J., 'The selection and performance of single-frequency sweep vibration tests', *Shock, Vibration and Associated Environments Bulletin*, n° 23, 1955, p. 93/101.

[CUR 71]   CURTIS A.J., TINLING N.G. and ABSTEIN H.T., 'Selection and performance of vibration tests', The Shock and Vibration Information Center, *SVM 8*. 1971.

[DEC 70]   DECLUE T.K., ARONE R.A. and DECKARD C.E., 'Multi-degree of freedom motion simulator systems for transportation environments', *The Shock and Vibration Bulletin*, n° 41, Part 3, Dec. 1970, p. 119/132.

[DEN 29]   DEN HARTOG J.P., 'Forced vibrations with Coulomb damping', *Univ. Pittsburgh Bull.*, Vol. 26, n° 1, Oct. 1929.

[DEN 30a]   DEN HARTOG J.P., 'Forced vibrations with combined viscous and Coulomb damping', *Philosophical Magazine*, Vol. 9, n° LIX, Suppl., May 1930, p. 801/817.

[DEN 30b]   DEN HARTOG J.P., 'Steady forced vibration as influenced by damping', by JACOBSEN L.S., *Discussion of Transactions of the ASME*, Vol. 52, Appl. Mech. Section, 1930, p. 178/180.

[DEN 56]   DEN HARTOG J.P., *Mechanical Vibrations*, McGraw-Hill Book Company, 1956.

[DEN 60]   DEN HARTOG J.P., *Vibrations Mécaniques*, Dunod, 1960.

[DEV 47]   'Development of NOL shock and vibration testing equipment', *The Shock and Vibration Bulletin*, n° 3, May 1947.

[DIM 61]   DIMENTBERG F.M., *Flexural Vibrations of Rotating Shafts*, Butterworths, London, 1961.

[DIT 67]   DITKIN V.A. and PRUDNIKOV A.P., *Formulaire Pour le Calcul Opérationnel*, Masson, 1967.

[DUB 59]   DUBLIN M., 'The nature of the vibration testing problem', *Shock, Vibration and Associated Environments Bulletin*, n° 27, Part IV, June 1959, p. 1/6.

[EAR 72]   EARLES S.W.E. and WILLIAMS E.J., 'A linearized analysis for frictionally damped systems', *Journal of Sound and Vibration*, Vol. 24, n° 4, 1972, p. 445/458.

[ELD 61]   ELDRED K., ROBERTS W.M. and WHITE R., 'Structural vibrations in space vehicles', *WADD Technical Report 61-62*, Dec. 1961.

[ENC 73]   *Encyclopédie Internationale des Sciences et des Techniques*, Presses de la Cité, 1973, Vol. 9, p. 539/543.

[FEL 59]   FELTNER C.E., 'Strain hysteresis, energy and fatigue fracture', *T.A.M. Report 146*, University of Illinois, Urbana, June 1959.

[FÖP 36]   FOPPL O., 'The practical importance of the damping capacity of metals, especially steels', *Journal of Iron and Steel*, Vol. 134, 1936, p. 393/455.

[FÖR 37]   FORSTER F., 'Ein neues Meßverfahren zur Bestimmung des Elastizitäts-moduls und der Dämpfung', *Zeitschrift für Metallkunde*, 29, Jahrgang, Heft 4, April 1937, p. 109/115.

[FOU 64]   FOUILLE A., *Electrotechnique*, Dunod, 1964.

[FRI 59]   FRIKE W. and KAMINSKY R.K., 'Application of reverberant and resonant chamber to acoustical testing of airborne components', *The Shock and Vibration Bulletin*, n° 27, Part II, June 1959, p. 216/225.

[FUN 58]   FUNG Y.C. and BARTON M.V., 'Some shock spectra characteristics and uses', *Journal of Applied Mechanics*, Vol. 35, Sept. 1958.

[GAB 69]   GABRIELSON V.K. and REESE R.T., 'Shock code user's manual – A computer code to solve the dynamic response of lumped-mass systems', *SCL-DR - 69 - 98*, November 1969.

[GAM 92]   GAM EG13, *Essais Généraux en Environnement des Matériels*, Annexe Générale Mécanique, DGA – Ministère de la Défense, 1992.

[GAN 85]   GANTENBEIN F. and LIVOLANT M., *Amortissement.. Génie Parasismique*, sous la direction de Victor Davidovici, Presses de l'Ecole Nationale des Ponts et Chaussées, 1985, p. 365/372.

[GER 61]   GERTEL M., *Specification of Laboratory Tests, Shock and Vibration Handbook*, Vol. 2, p. 24 - 1/24 - 34, HARRIS C.M. and CREDE C.E., McGraw-Hill Book Company, 1961.

[GOO 76]   GOODMAN L.E., *Material Damping and Slip Damping, Shock and Vibration Handbook*, Vol. 36, McGraw Hill Book Company, 1976.

[GUI 63]   GUILLIEN R., *Electronique*, Presses Universitaires de France, Vol. 1, 1963.

[GUR 59]   GURTIN M., 'Vibration analysis of discrete mass systems', General Engineering Laboratory, *G.E. Report n° 59GL75*, 15 March 1959.

[HAB 68]   HABERMAN C.M., *Vibration Analysis*, C.E. Merril Publishing Company, Columbus, Ohio, 1968.

[HAL 75]   HALLAM M.G., HEAF N.J. and WOOTTON L.R., 'Dynamics of marine structures: methods of calculating the dynamic response of fixed structures subject to wave and current action', CIRIA Underwater Engineering Group, ATKINS Research and Development, *Report UR 8*, Oct. 1975.

[HAL 78]   HALLAM M.G., HEAF N.J. and WOOTTON L.R., 'Dynamics of marine structures: Methods of calculating the dynamic response of fixed structures subject to wave and current action', CIRIA Underwater Engineering Group, *Report UR 8*, Oct. 1978.

[HAU 65]   HAUGEN E.B., 'Statistical strength properties of common metal alloys', North American Aviation Inc., Space and Information Systems Division, *SID 65 - 1274*, 30 Oct. 1965.

[HAW 64]   HAWKES P.E., 'Response of a single-degree-of-freedom system to exponential sweep rates', *Shock, Vibration and Associated Environments*, n° 33, Part 2, Feb. 1964, p 296/304. (Or *Lockheed Missiles and Space Company Structures Report LMSC A 362881 - SS/690*, 12 Nov. 1963).

[HAY 72]   HAY J.A., 'Experimentally determined damping factors', *Symposium on Acoustic Fatigue, AGARD CP 113*, Sept. 1972, p. 12-1/12-15.

[HLA 69]   HLADIK J., *La Transformation de Laplace à Plusieurs Variables*, Masson, 1969.

[HOB 76]   HOBAICA E.C. and SWEET G., 'Behaviour of elastomeric materials under dynamic loads', *The Shock and Vibration Digest*, Vol. 8, n° 3, March 1976, p. 77/78.

[HOK 48]   HOK G., 'Response of linear resonant systems to excitation of a frequency varying linearly with time', *Journal of Applied Physics*, Vol. 19, 1948, p. 242/250.

[HOP 04]   HOPKINSON B., 'The effects of momentary stresses in metals', *Proceedings of the Royal Society of London*, Vol. 74, 1904–5, p. 717/735.

[HOP 12]   HOPKINSON B. and TREVOR-WILLIAMS G., 'The elastic hysteresis of steel'. *Proceedings of the Royal Society of London*, Series A, Vol. 87, 1912, p. 502.

[IMP 47]   'Impressions of the shock and vibration tour', Naval Research Laboratory, *The Shock and Vibration Bulletin*, n° 2, March 1947.

[JAC 30]   JACOBSEN L.S., 'Steady forced vibration as influenced by damping', *Transactions of the ASME 52*, Appl. Mech. Section, 1930, p. 169/178.

[JAC 58]   JACOBSEN L.S. and AYRE R.S., *Engineering Vibrations*, McGraw-Hill Book Company, Inc., 1958.

[JEN 59]   JENSEN J.W., 'Damping capacity: Its measurement and significance', *Report of Investigations 5441*, US Bureau of Mines, Washington, 1959.

[JON 69]   JONES D.I.G., HENDERSON J.P. and NASHIF A.D., 'Reduction of vibrations in aerospace structures by additive damping', *The Shock and Vibration Bulletin*, n° 40, Part 5, 1969, p. 1/18.

[JON 70]   JONES D.I.G., HENDERSON J.P. and NASHIF A.D., 'Use of damping to reduce vibration induced failures in aerospace systems', *Proceedings of the Air Force Conference on Fatigue and Fracture of Aircraft Structures and Materials*. Miami Beach, 15–18 Dec. 1969, *AFFDL TR70-144*, 1970, p. 503/519.

[KAR 40]   KARMAN T.V. and BIOT M.A., *Mathematical Methods in Engineering*. McGraw-Hill Book Company, Inc., 1940.

[KAR 50]   KARMAN T.V. and DUWEZ P.E., 'The propagation of plastic deformation in solids', *Journal of Applied Physics*, Vol. 21, 1950, p. 987.

[KAY 77]   KAYANICKUPURATHU J.T., 'Response of a hardening spring oscillator to random excitation', *The Shock and Vibration Bulletin*, n°47, Part 2, 1977, p. 5/9.

[KEN 47]   KENNEDY C.C. and PANCU C.D.P., 'Use of vectors in vibration measurement and analysis', *Journal of the Aeronautical Sciences*, Vol. 14, 1947, p. 603/625.

[KEV 71]   KEVORKIAN J., 'Passage through resonance for a one-dimensional oscillator with slowly varying frequency', *SIAM Journal of Applied Mathematics*, Vol. 20, n° 3, May 1971.

[KHA 57]   KHARKEVTICH A.A., *Les Spectres et L'analyse*, Editions URSS, Moscow, 1957.

[KIM 24]   KIMBALL A.L., 'Internal friction theory of shaft whirling', *General Electric Review*, Vol. 27, April 1924, p. 244.

[KIM 26]   KIMBALL A.L. and LOVELL D.E., 'Internal friction in solids', *Transactions of the ASME*, Vol. 48, 1926, p. 479/500.

[KIM 27]   KIMBALL A.L. and LOVELL D.E., 'Internal friction in solids', *Physical Review*, Vol. 30, Dec. 1927, p. 948/959.

[KIM 29]   KIMBALL A.L., 'Vibration damping, including the case of solid friction', *ASME*, *APM-51-21*, 1929.

[KLE 71a]   KLESNIL M., LUKAS P. and RYS P., Inst. of Phys. Met., *Czech Academy of Sciences Report*, Brno, 1971.

[KLE 71b]   KLEE B.J., *Design for Vibration and Shock Environments*, Tustin Institute of Technology, Santa Barbara, California. 1971.

[LAL 75]   LALANNE C., 'La simulation des environnements de choc mécanique', *Rapport CEA-R - 4682*, Vols. 1 and 2, 1975.

[LAL 80]   LALANNE M., BERTHIER P. and DER HAGOPIAN J., *Mécanique des Vibrations Linéaires*, Masson, 1980.

[LAL 82]   LALANNE C., 'Les vibrations sinusoïdales à fréquence balayée', *CESTA/EX* n° 803, 8 June 1982.

[LAL 95a]   LALANNE C., 'Analyse des vibrations aléatoires', *CESTA/DQS DO 60*, 10 May 1995.

[LAL 95b]   LALANNE C., 'Vibrations aléatoires – Dommage par fatigue subi par un système mécanique à un degré de liberté', *CESTA/DT/EX DO 1019*, 20 Jan. 1995.

[LAL 96]   LALANNE C., 'Vibrations mécaniques', *CESTA/DQS DO 76*, 22 May 1996.

[LAN]   LANDAU L. and LIFCHITZ E., *Mécanique, Physique Théorique*, Editions de la Paix, Vol. 1.

[LAZ 50]   LAZAN B.J., 'A study with new equipment of the effects of fatigue stress on the damping capacity and elasticity of mild steel', *Transactions of the ASME*, Vol. 42, 1950, p. 499/558.

[LAZ 53]   LAZAN B.J., 'Effect of damping constants and stress distribution on the resonance response of members', *Journal of Applied Mechanics. Transactions of the ASME*, Vol. 20, 1953, p. 201/209.

[LAZ 68]   LAZAN B.J., *Damping of Materials and Members in Structural Mechanics*, Pergamon Press, 1968.

[LEV 60]   LEVITAN E.S., 'Forced oscillation of a spring–mass system having combined Coulomb and viscous damping', *Journal of the Acoustical Society of America*, Vol. 32, n° 10, Oct. 1960, p. 1265/1269.

[LEV 76]   LEVY S. and WILKINSON J.P.D., *The Component Element Method in Dynamics*, McGraw-Hill Book Company, 1976.

[LEW 32]   LEWIS F.M., 'Vibration during acceleration through a critical speed', *Transactions of the ASME, APM 54 – 24*, 1932, p. 253/261

[LOR 70]   LORENZO C.F., 'Variable-sweep-rate testing: a technique to improve the quality and acquisition of frequency response and vibration data', *NASA Technical Note D-7022*, Dec. 1970.

[MAB 84]   MABON L., PRUHLIERE J.P., RENOU C. and LEJUEZ W., 'Modèle mathématique d'un véhicule', *ASTE. IXe Journées Scientifiques et Techniques*, Paris, 6–8 March 1984, p. 153/161.

[MAC 58]   MACDUFF J.N. and CURRERI J.R., *Vibration Control*, McGraw-Hill Book Company, Inc., 1958.

[MAZ 66]   MAZET R., *Mécanique Vibratoire*. Dunod, 1966.

[MEI 67]   MEIROVITCH L., *Analytical methods in vibrations*, The Macmillan Company. New York, 1967.

[MIN 45]   MINDLIN R.C., 'Dynamics of package cushioning', *Bell System Technical Journal*, Vol. 24, July–Oct. 1945, p. 353/461.

[MOR 53]   MORROW C.T. and MUCHMORE R.B., 'Simulation of continuous spectra by line spectra in vibration testing', *The Shock and Vibration Bulletin*, n° 21, Nov. 1953.

[MOR 63a]   MORLEY A.W. and BRYCE W.D., 'Natural vibration with damping force proportional to a power of the velocity', *Journal of the Royal Aeronautical Society*, Vol. 67, June 1963, p. 381/385.

[MOR 63b]   MORROW C.T., *Shock and Vibration Engineering*, John Wiley and Sons Inc., Vol. 1, 1963.

[MOR 65]   MORSE R.E., 'The relationship between a logarithmically swept excitation and the build-up of steady-state resonant response', *The Shock and Vibration Bulletin*, n° 35, Part II, 1965, p. 231/262.

[MOR 76]   MORROW T., 'Environmental specifications and testing', in HARRIS C.M. and CREDE C.E (Eds), *Shock and Vibration Handbook*, 2nd ed., p. 24 - 1/24 - 13, McGraw-Hill Book Company, 1976.

[MUR 64]   MURFIN W.B., 'Dynamics of mechanical systems', Sandia National Labs, *RPT SC-TM 640931*, Aug. 1964.

[MUS 68]   MUSTER D., 'International standardization in mechanical vibration and shock', *Journal of Environmental Sciences*, Vol. 11, n° 4, Aug. 1968, p. 8/12.

[MYK 52]   MYKLESTAD N.O., 'The concept of complex damping', *Journal of Applied Mechanics*, Vol. 19, 1952, p. 284/286.

[NEL 80]   NELSON F.C. and GREIF R., 'Damping models and their use in computer programs', University Press of Virginia, *Structural Mechanics Software Series*, Vol. 3, 1980, p. 307/337.

[OLS 57]   OLSON M.W., 'A narrow-band-random-vibration test', *The Shock and Vibration Bulletin*, n° 25, Part I, Dec. 1957, p. 110.

[PAI 59]   PAINTER G.W., 'Dry-friction damped isolators', *Prod. Eng.*, Vol. 30, n° 31, 3 Aug. 1959, p. 48/51.

[PAR 61]   PARKER A.V., 'Response of a vibrating system to several types of time-varying frequency variations', *Shock, Vibration and Associated Environments Bulletin*, n° 29, Part IV, June 1961, p. 197/217.

[PEN 65]   PENNINGTON D., *Piezoelectric Accelerometer Manual*, Endevco Corporation, Pasadena, California, 1965.

[PIE 64]   PIERSOL A.G., 'The measurement and interpretation of ordinary power spectra for vibration problems', *NASA - CR 90*, 1964.

[PIM 62]   PIMONOW L., *Vibrations en Régime Transitoire*, Dunod, 1962.

[PLU 59]   PLUNKETT R., 'Measurement of damping', in RUZICKA J. (Ed), *Structural Damping, ASME*, Section Five, Dec. 1959, p. 117/131.

[POT 48]   POTTER E.V., 'Damping capacity of metals', USBRMI, Wash., *R. of I. 4194*, March 1948.

[PUS 77]   PUSEY H.C., 'An historical view of dynamic testing', *Journal of Environmental Sciences*, Sept./Oct. 1977, p. 9/14.

[QUE 65]   *Quelques Formes Modernes de Mathématiques*, Publications de l'OCDE, Nov. 1965.

[REE 60]   REED W.H., HALL A.W. and BARKER L.E., 'Analog techniques for measuring the frequency response of linear physical systems excited by frequency sweep inputs', *NASA TN D 508*, 1960.

[REI 56]   REID T.J., 'Free vibration and hysteretic damping', *Journal of the Royal Aeronautical Society*, Vol. 60, 1956, p. 283.

[RID 69]   RIDLER K.D. and BLADER F.B., 'Errors in the use of shock spectra', *Environmental Engineering*, July 1969, p. 7/16.

[ROO 82]   ROONEY G.T. and DERAVI P., 'Coulomb friction in mechanism sliding joints'. *Mechanism and Machine Theory*, Vol. 17, n° 3, 1982, p. 207/211.

[RUB 64]   RUBIN S., 'Introduction to dynamics', *Proceedings of the IES*, 1964, p. 3/7.

[RUZ 57]   RUZICKA J.E., 'Forced vibration in systems with elastically supported dampers', *Masters Thesis*, MIT, Cambridge, Mass., June 1957.

[RUZ 71]   RUZICKA J.E. and DERBY T.F., 'Influence of damping in vibration isolation', The Shock and Vibration Information Center, USDD, *SVM-7*, 1971.

[SAL 71]   SALLES F., *Initiation au Calcul Opérationnel et à Ses Applications Techniques*, Dunod, 1971.

[SCA 63]   SCANLAN R.H. and MENDELSON A., 'Structural damping', *AIAA Journal*, Vol. 1, n° 4, April 1963, p. 938/939.

[SKI 66]   SKINGLE C.W., 'A method for analysing the response of a resonant system to a rapid frequency sweep input', *RAE Technical Report 66379*, Dec. 1966.

[SMA 85]   SMALLWOOD D.O., 'Shock testing by using digital control', *SANDIA 85 - 0352 J*, 1985.

[SNO 68]   SNOWDON J.C., *Vibration and Shock in Damped Mechanical Systems*, John Wiley and Sons, Inc., 1968.

[SOR 49]   SOROKA W.W., 'Note on the relations between viscous and structural damping coefficients', *Journal of the Aeronautical Sciences*, Vol. 16, July 1949, p. 409/410.

[SPE 61]   SPENCE H.R. and LUHRS H.N., 'Peak criterion in random vs sine vibration testing', *Journal of the Acoustical Society of America*, Vol. 33, n° 5, May 1961. p. 652/654.

[SPE 62]   SPENCE H.R. and LUHRS H.N.. 'Structural fatigue under combined random and swept sinusoidal vibration', *Journal of the Acoustical Society of America*, Vol. 34, n° 8, Aug. 1962, p. 1098/1101.

[STA 53]   STANTON L.R. and THOMSON F.C., 'A note on the damping charistics of some magnesium and aluminum alloys', *Journal of the Institute of Metals*, Vol. 69, Part 1, 1953, p. 29.

[STE 73]   STEINBERG D.S., *Vibrations Analysis for Electronic Equipment*, John Wiley and Sons, 1973.

[STE 78]   STEINBERG D.S., 'Quick way to predict random vibration failures', *Machine Design*, Vol. 50, n° 8, 6 April 1978, p. 188/191.

[SUN 75]   SUNG L.C., 'An approximate solution for sweep frequency vibration problems', *PhD Dissertation*, Ohio State University, 1975.

[SUN 80]   SUNG L. and STEVENS K.K., 'Response of linear discrete and continuous systems to variable frequency sinusoidal excitations', *Journal of Sound and Vibration*, Vol. 71, n° 4, 1980, p. 497/509.

[SUT 68]   SUTHERLAND L.C., 'Fourier spectra and shock spectra for simple undamped systems', *NASA-CR 98417*, Oct. 1968.

[SUZ 78a]   SUZUKI S.I., 'Dynamic behaviour of a beam subjected to a force of time-dependent frequency (continued)', *Journal of Sound and Vibration*, Vol. 60, n° 3, 1978, p. 417/422.

[SUZ 78b]   SUZUKI S.I., 'Dynamic behaviour of a beam subjected to a force of time-dependent frequency', *Journal of Sound and Vibration*, Vol. 57, n° 1, 1978, p. 59/64.

[SUZ 79]   SUZUKI S.I., 'Dynamic behaviour of a beam subjected to a force of time-dependent frequency (effects of solid viscosity and rotatory inertia)', *Journal of Sound and Vibration*, Vol. 62, n° 2, 1979, p. 157/164.

[TAY 46]   TAYLOR G.I., 'The testing of materials at high rates of loading', *Journal of the Institute of Civil Engineers*, Vol. 26, 1946, p. 486/519.

[TAY 75]   TAYLOR H.R., 'A study of instrumentation systems for the dynamic analysis of machine tools', *PhD Thesis*, University of Manchester, 1975.

[TAY 77]   TAYLOR H.R., 'A comparison of methods for measuring the frequency response of mechanical structures with particular reference to machine tools', *Proceedings of the Institute of Mechanical Engineers*, Vol. 191, 1977, p. 257/270.

[THO 65a]   THOMSON W.T., *Vibration Theory and Applications*, Prentice Hall, Inc., 1965.

[THO 65b]   THOMSON, Sir W. (Lord Kelvin), 'On the elasticity and viscosity of metals', *Proceedings of the Royal Society of London*, Vol. 14, 1865, p. 289.

[THU 71]   THUREAU P. and LECLER D., *Vibrations – Régimes Linéaires*, Technologie et Université, Dunod, 1971.

[TRU 70]   TRULL R.V., 'Sweep speed effects in resonant systems', *The Shock and Vibration Bulletin*, Vol. 41, Part 4, Dec. 1970, p. 95/98.

[TRU 95]   TRULL R.V., ZIMMERMANN R.E. and STEIN P.K., 'Sweep speed effects in resonant systems: A unified approach, Parts I, II and III', *Proceedings of the 66th Shock and Vibration Symposium*, Vol. II, 1995, p. 115/146.

[TUR 54]   TURBOWITCH I.T., 'On the errors in measurements of frequency characteristics by the method of frequency modulation', *Radiotekhnika*, Vol. 9, 1954, p. 31/35.

[TUS 72]   TUSTIN W., *Environmental Vibration and Shock: Testing, measurement, analysis and calibration*, Tustin Institute of Technology, Santa Barbara, California, 1972.

[UNG 73]   UNGAR E.E., 'The status of engineering knowledge concerning the damping of built-up structures', *Journal of Sound and Vibration*, Vol. 26, n° 1, 1973, p. 141/154.

[VAN 57]   VAN BOMMEL P., 'A simple mass-spring-system with dry damping subjected to harmonic vibrations', *De Ingenieur*, Nederl., Vol. 69, n° 10, 1957, p. w37/w44.

[VAN 58]   VAN BOMMEL P., 'An oscillating system with a single mass with dry frictional damping subjected to harmonic vibrations', *Bull. Int. R. Cong. XXXV*, n° 1, Jan. 1958, p. 61/72.

[VER 67]   VERNON J.B., *Linear Vibration Theory*, John Wiley and Sons, Inc., 1967.

[VOL 65]   VOLTERRA E. and ZACHMANOGLOU E.C., *Dynamics of Vibrations*, Charles E. Merril Books, Inc., 1965.

[WEG 35]   WEGEL R.L. and WALTHER H., 'Internal dissipation in solids for small cyclic strains', *Physics*, Vol. 6, 1935, p. 141/157.

[WHI 72]   WHITE R.G., 'Spectrally shaped transient forcing functions for frequency response testing', *Journal of Sound and Vibration*, Vol. 23, n° 3, 1972, p. 307/318.

[WHI 82]   WHITE R.G. and PINNINGTON R.J., 'Practical application of the rapid frequency sweep technique for structural frequency response measurement', *Aeronautical Journal*, n° 964, May 1982, p. 179/199.

[ZEN 40]   ZENER C., 'Internal friction in solids', *Proceedings of the Physical Society of London*, Vol. 52, Part 1, n° 289, 1940, p. 152.

Index

Printed and bound by CPI Group (UK) Ltd, Croydon, CR0 4YY

23/10/2024

01777670-0005